Recombinant DNA and Medical Genetics

BIRTH DEFECTS INSTITUTE SYMPOSIA

Ernest B. Hook, Dwight T. Janerich, and Ian H. Porter, editors: MONITORING, BIRTH DEFECTS, AND ENVIRONMENT: The Problem of Surveillance, 1971

Ian H. Porter and Richard G. Skalko, editors: HEREDITY AND SOCIETY, 1972

Dwight T. Janerich, Richard G. Skalko, and Ian H. Porter, editors: CONGENITAL DEFECTS: New Directions in Research, 1973

Hilaire J. Meuwissen, Richard J. Pickering, Bernard Pollara, and Ian H. Porter, editors: COMBINED IMMUNODEFICIENCY DISEASE AND ADENOSINE DEAMINASE DEFICIENCY: A Molecular Defect, 1975

Sally Kelly, Ernest B. Hook, Dwight T. Janerich, and Ian H. Porter, editors: BIRTH DEFECTS: Risks and Consequences, 1976

Ernest B. Hook and Ian H. Porter, editors: POPULATION CYTOGENETICS: Studies in Humans, 1977

H. Lawrence Vallet and Ian H. Porter, editors: GENETIC MECHANISMS OF SEXUAL DEVELOPMENT, 1979

Ian H. Porter and Ernest B. Hook, editors: SERVICE AND EDUCATION IN MEDICAL GENETICS, 1979

Bernard Pollara, Richard J. Pickering, Hilaire J. Meuwissen, and Ian H. Porter, editors: INBORN ERRORS OF SPECIFIC IMMUNITY, 1979

Ian H. Porter and Ernest B. Hook, editors: HUMAN EMBRYONIC AND FETAL DEATH, 1980

Ernest B. Hook and Ian H. Porter, editors: POPULATION AND BIOLOGICAL ASPECTS OF HUMAN MUTATION, 1981

Ann M. Willey, Thomas P. Carter, Sally Kelly, and Ian H. Porter, editors: CLINICAL GENETICS: Problems in Diagnosis and Counseling, 1982

Anne Messer and Ian H. Porter, editors: RECOMBINANT DNA AND MEDICAL GENETICS, 1983

Recombinant DNA and Medical Genetics

Edited by

Anne Messer **Ian H. Porter**

Birth Defects Institute
Center for Laboratories and Research
New York State Department of Health
Albany, New York

1983

ACADEMIC PRESS
A Subsidiary of Harcourt Brace Jovanovich, Publishers
New York London
Paris San Diego San Francisco São Paulo Sydney Tokyo Toronto

Proceedings of the Thirteenth Annual
New York State Health Department
Birth Defects Symposium

Academic Press Rapid Manuscript Reproduction

ACADEMIC PRESS, INC.
111 Fifth Avenue, New York, New York 10003

United Kingdom Edition published by
ACADEMIC PRESS, INC. (LONDON) LTD.
24/28 Oval Road, London NW1 7DX

Library of Congress Cataloging in Publication Data

Main entry under title:

Recombinant DNA and medical genetics.

(Birth Defects Institute symposia ; #13)
Proceedings of a symposium held Nov. 15-16, 1982, Albany, N.Y. and sponsored by the Birth Defects Institute of the New York State Dept. of Health.
Includes index.
1. Medical genetics--Congresses. 2. Recombinant DNA--Congresses. I. Messer, Anne. II. Porter, Ian H. III. New York (State). Birth Defects Institute. IV. Series. [DNLM: 1. Genetic intervention--Congresses. 2. Recombinant DNA--Congresses. QH 438.7 R3097 1982]
RB155.R337 1983 616'.042 83-15668
ISBN 0-12-492220-1

PRINTED IN THE UNITED STATES OF AMERICA

83 84 85 86 9 8 7 6 5 4 3 2 1

Contents

Abstracts

Contributors

Numbers in parentheses indicate the pages on which the authors' contributions begin.

Kenneth Abel (200), *Roswell Park Memorial Institute, New York State Department of Health, Buffalo, New York 14263*

Jeff Aldridge (35), *Genetics Division and Mental Retardation Center, Children's Hospital Medical Center and Department of Pediatrics, Harvard Medical School, Boston, Massachusetts 02138*

Stylianos E. Antonarakis (135), *Department of Pediatrics, The Johns Hopkins University School of Medicine, Baltimore, Maryland 21218*

Margaret M. Aronson (193), *Institute for Medical Research, Camden, New Jersey 08103*

James A. Barbosa (202), *Yale University, New Haven, Connecticut 06520*

David Barker (73), *Department of Cellular, Viral, and Molecular Biology and Howard Hughes Medical Institute, University of Utah Medical School, Salt Lake City, Utah 84112*

Marlene Belfort (204), *Center for Laboratories and Research, New York State Department of Health, Albany, New York 12237*

Jon Berkowitz (73), *Department of Cellular, Viral, and Molecular Biology and Howard Hughes Medical Institute, University of Utah Medical School, Salt Lake City, Utah 84112*

Corinne D. Boehm (135), *Department of Pediatrics, The Johns Hopkins University School of Medicine, Baltimore, Maryland 21218*

William L. Boyko (201), *Children's Hospital Research Foundation, Cincinnati, Ohio 45229*

Robert D. Burk (195), *Howard Hughes Medical Institute, The Johns Hopkins University School of Medicine, Baltimore, Maryland 21218*

R. Daniel Camerini-Otero (105), *Genetics and Biochemistry Branch, National Institute of Arthritis, Diabetes, and Digestive and Kidney Diseases, National Institutes of Health, Bethesda, Maryland 20205*

Web Cavenee (73), *Department of Cellular, Viral, and Molecular Biology and Howard Hughes Medical Institute, University of Utah Medical School, Salt Lake City, Utah 84112*

Verne M. Chapman (197, 199), *Roswell Park Memorial Institute, New York State Department of Health, Buffalo, New York 14263*

Tu-chen Cheng (135), *Department of Pediatrics, The Johns Hopkins University School of Medicine, Baltimore, Maryland 21218*

Lewis L. Coriell (193), *Institute for Medical Research, Camden, New Jersey 08103*

Robert DeMars (117), *Laboratory of Genetics and Department of Human Oncology, University of Wisconsin, Madison, Wisconsin 53706*

Bérengère de Martinville (21), *Department of Human Genetics, Yale University School of Medicine, New Haven, Connecticut 06520*

Douglas Dickinson (200), *Roswell Park Memorial Institute, New York State Department of Health, Buffalo, New York 14263*

Dennis Drayna (73), *Department of Cellular, Viral, and Molecular Biology and Howard Hughes Medical Institute, University of Utah Medical School, Salt Lake City, Utah 84112*

Eva M. Eicher (57), *The Jackson Laboratory, Bar Harbor, Maine 04609*

Rosemary W. Elliott (196), *Roswell Park Memorial Institute, New York State Department of Health, Buffalo, New York 14263*

Glen A. Evans (105), *Laboratory of Molecular Genetics, National Institute of Child Health and Human Development, National Institutes of Health, Bethesda, Maryland 20205*

A. Ferguson-Smith (202), *Yale University, New Haven, Connecticut 06520*

Stuart G. Fischer (157, 189), *Center for Biological Macromolecules, Department of Biological Sciences, New York State University at Albany, Albany, New York 12203*

Lorraine Flaherty (1), *Center for Laboratories and Research, New York State Department of Health, Albany, New York 12237*

Lesley M. Forrester (197, 199), *Roswell Park Memorial Institute, New York State Department of Health, Buffalo, New York 14263*

Uta Francke (21), *Department of Human Genetics, Yale University School of Medicine, New Haven, Connecticut 06520*

Patricia M. Gallagher (201), *Children's Hospital Research Foundation, Cincinnati, Ohio 45229*

Roger E. Ganschow (201), *Children's Hospital Research Foundation, Cincinnati, Ohio 45229*

Arthur E. Greene (193), *Institute for Medical Research, Camden, New Jersey 08103*

Dian E. Grogan (201), *Children's Hospital Research Foundation, Cincinnati, Ohio 45229*

Kenneth W. Gross (200), *Roswell Park Memorial Institute, New York State Department of Health, Buffalo, New York 14263*

Nicholas D. Hastie (197, 199), *Roswell Park Memorial Institute, New York State Department of Health, Buffalo, New York 14263*

Ronald C. Herman (203), *Center for Laboratories and Research, New York State Department of Health, Albany, New York 12237*

Tom Holm (73), *Department of Cellular, Viral, and Molecular Biology and Howard Hughes Medical Institute, University of Utah Medical School, Salt Lake City, Utah 84112*

Lee Hood (97), *Division of Biology, California Institute of Technology, Pasadena, California 91125*

Henry Huang (97), *Department of Microbiology and Immunology, Washington University School of Medicine, St. Louis, Missouri 63130*

Cecile Huerre (191), *Institut de Pathologie et Biologie Cellulaire et Moléculaire and Institut de Progénèse, Paris, France*

Ethylin W. Jabs (49), *Department of Pediatrics, The Johns Hopkins University School of Medicine, Baltimore, Maryland 21218*

Claudine Junien (191), *Institut de Pathologie et Biologie Cellulaire et Moléculaire and Institut de Progénèse, Paris, France*

Michael E. Kamark (202), *Yale University, New Haven, Connecticut 06520*

Haig H. Kazazian, Jr. (135), *Department of Pediatrics, The Johns Hopkins University School of Medicine, Baltimore, Maryland 21218*

John L. Knopf (200), *Roswell Park Memorial Institute, New York State Department of Health, Buffalo, New York 14263*

Louis M. Kunkel (35), *Genetics Division and Mental Retardation Center, Children's Hospital Medical Center and Department of Pediatrics, Harvard Medical School, Boston, Massachusetts 02138*

Marc Lalande (35), *Genetics Division and Mental Retardation Center, Children's Hospital Medical Center and Department of Pediatrics, Harvard Medical School, Boston, Massachusetts 02138*

Samuel A. Latt (35), *Genetics Division and Mental Retardation Center, Children's Hospital Medical Center and Department of Pediatrics, Harvard Medical School, Boston, Massachusetts 02138*

Robin Leach (73), *Department of Cellular, Viral, and Molecular Biology and Howard Hughes Medical Institute, University of Utah Medical School, Salt Lake City, Utah 84112*

Mark Leppert (73), *Department of Cellular, Viral, and Molecular Biology and Howard Hughes Medical Institute, University of Utah Medical School, Salt Lake City, Utah 84112*

Leonard S. Lerman (157, 189), *Center for Biological Macromolecules, Department of Biological Sciences, New York State University at Albany, Albany, New York 12203*

Nadya Lumelsky (157), *Center for Biological Macromolecules, Department of Biological Sciences, New York State University at Albany, Albany, New York 12203*

Frank Maley (204), *Center for Laboratories and Research, New York State Department of Health, Albany, New York 12237*

Gladys Maley (204), *Center for Laboratories and Research, New York State Department of Health, Albany, New York 12237*

Elizabeth Mann (196), *Roswell Park Memorial Institute, New York State Department of Health, Buffalo, New York 14263*

David Margulies (105), *Laboratory of Molecular Genetics, National Institute of Child Health and Human Development, National Institutes of Health, Bethesda, Maryland 20205*

Barbara R. Migeon (49), *Deparment of Pediatrics, The Johns Hopkins University School of Medicine, Baltimore, Maryland 21218*

Alan Moelleken (204), *Center for Laboratories and Research, New York State Department of Health, Albany, New York 12237*

Richard A. Mulivor (193), *Institute for Medical Research, Camden, New Jersey 08103*

Warren W. Nichols (193), *Institute for Medical Research, Camden, New Jersey 08103*

Stuart H. Orkin (145), *Division of Hematology and Oncology, Children's Hospital Medical Center and the Sidney Farber Cancer Institute and Department of Pediatrics, Harvard Medical School, Boston, Massachusetts 02138*

Harry T. Orr (117), *Immunobiology Research Center, Department of Laboratory Medicine/Pathology, University of Minnesota, Minneapolis, Minnesota 55455*

Keiko Ozato (105), *Pregnancy Research Branch, National Institute of Child Health and Human Development, National Institutes of Health, Bethesda, Maryland 20205*

Robbin Palmer (201), *Children's Hospital Research Foundation, Cincinnati, Ohio 45229*

John A. Phillips III (9), *Department of Pediatrics, The Johns Hopkins University School of Medicine, Baltimore, Maryland 21218*

Sandra J. Phillips (57), *The Jackson Laboratory, Bar Harbor, Maine 04609*

Nina Piccini (200), *Roswell Park Memorial Institute, New York State Department of Health, Buffalo, New York 14263*

Marie-Odile Rethore (191), *Institut de Pathologie et Biologie Cellulaire et Moléculaire and Institut de Progénèse, Paris, France*

Beverly Richards-Smith (196), *Roswell Park Memorial Institute, New York State Department of Health, Buffalo, New York 14263*

Janet Rossant (199), *Brock University, St. Catharines, Ontario, Canada*

Frank H. Ruddle (202), *Yale University, New Haven, Connecticut 06520*

Janet P. Sanford (197), *Roswell Park Memorial Institute, New York State Department of Health, Buffalo, New York 14263*

J. G. Seidman (105), *Department of Genetics, Harvard Medical School, Boston, Massachusetts 02138*

Thomas B. Shows (79), *Department of Human Genetics, Roswell Park Memorial Institute, New York State Department of Health, Buffalo, New York 14263*

Kirby D. Smith (195), *Howard Hughes Medical Institute, The Johns Hopkins University School of Medicine, Baltimore, Maryland 21218*

Judy Stamberg (195), *Howard Hughes Medical Institute, The Johns Hopkins University School of Medicine, Baltimore, Maryland 21218*

Umadevi Tantravahi (35), *Genetics Division and Mental Retardation Center, Children's Hospital Medical Center and Department of Pediatrics, Harvard Medical School, Boston, Massachusetts 02138*

James V. Tricoli (79), *Department of Human Genetics, Roswell Park Memorial Institute, New York State Department of Health, Buffalo, New York 14263*

Pamela G. Waber (135), *Department of Pediatrics, The Johns Hopkins University School of Medicine, Baltimore, Maryland 21218*

Linda L. Washburn (57), *The Jackson Laboratory, Bar Harbor, Maine 04609*

Ray White (73), *Department of Cellular, Viral, and Molecular Biology and Howard Hughes Medical Institute, University of Utah Medical School, Salt Lake City, Utah 84112*

Ann Willey,* *Birth Defects Institute, Center for Laboratories and Research, New York State Department of Health, Albany, New York 12237*

Stanley F. Wolf (49), *Department of Pediatrics, The Johns Hopkins University School of Medicine, Baltimore, Maryland 21218*

Keith E. Young (195), *Howard Hughes Medical Institute, The Johns Hopkins University School of Medicine, Baltimore, Maryland 21218*

Bernhard U. Zabel (79), *Department of Human Genetics, Roswell Park Memorial Institute, New York State Department of Health, Buffalo, New York 14263*

*Session moderator.

Preface

Thirty years ago this month Watson and Crick published a paper (*Nature [London]* **171,** 737 [1953]) describing the nature of the double helix of DNA and noting the copying mechanism that specific base pairing allowed. Last week a "routine" journal paper (*N. Engl. J. Med.* **308,** 1054 [1983]) reported using DNA polymorphisms for prenatal diagnosis of 95 pregnancies at risk for sickle-cell disease or β-thalassemia. Thus, we can trace the "New Genetics," developing complex techniques that enable us to isolate or clone a single, specific DNA fragment carrying a gene of special interest. It is the great power of this constellation of techniques that is revolutionizing medical genetics. Both the sophistication and ease with which specific hereditary diseases can be diagnosed are rapidly changing. In addition, our knowledge of some of the basic processes in the functioning of the gene should lead to further insights concerning the nature of the mutations and diseases they cause and possibly to innovative new forms of treatment. This volume contains the proceedings (papers and poster abstracts) of Birth Defects Symposium XIII, "Recombinant DNA and Medical Genetics," held November 15 and 16, 1982, at the Albany Hilton Hotel, Albany, New York. Nearly 300 participants heard talks, viewed poster presentations, saw examples of some of the latest commercial aids for using this technology, and discussed both theoretical and practical issues, formally and informally.

An introduction to molecular genetics has been provided for those readers who wish to review terminology and techniques. A variety of work examining a range of known mutations (e.g., growth hormone deficiencies and hemoglobinopathies), polymorphisms (mostly of unknown function), and specific chromosomal probes, both *in vitro* and *in situ,* is presented. Based on these kinds of studies, linkage analyses may soon be available for a number of hereditary diseases. The next two sets of papers represent the two areas in which recombinant DNA studies have provided the most concrete and dramatic advances to date. Basic research on the genes that specify antibodies and histocompatibility has elucidated some of the somatic mechanisms by which the specificity and diversity required of the immune system is generated. Studies of the hemoglobinopathies offer detailed descriptions of nucleotide sequences of mutant genes of substantial clinical interest, which should allow both precise prenatal diagnosis using small amounts of fetal DNA and a description of the mechanisms and levels of control of globin gene expression.

The final paper, by Lerman *et al.*, gives us a glimpse of what the future may hold technically in this field. Increasingly sophisticated physical methods of looking at small changes in specific DNA sequences will greatly enhance both diagnostic procedures and our rudimentary knowledge of how genes actually work. The eloquent introduction to this paper also serves as a summary and a current biochemical (and, to some extent, philosophical) overview of the genome.

As part of an ongoing discussion of how the work that was presented could actually be used clinically, David Houseman (of the Massachusetts Institute of Technology and Integrated Genetics) presented an after-dinner talk on the role of the private sector in genetic screening using recombinant DNA methods. His candid discussion and personal anecdotes of the experiences of an academician dealing with the world of business were much appreciated.

Finally, a technical note: Given the variety of organisms, genes, cloning vectors, and enzymes cited in these manuscripts, nomenclature turned into an editor's nightmare. We tried to follow the guidelines of the International System for Human Gene Nomenclature (T. B. Shows *et al.*, *Cytogenet. Cell Genet.* **25,** 96–116 [1979]) for human genes; the rules of the Committee on Standardized Genetic Nomenclature for Mice (*in* "Genetic Variants and Strains of the Laboratory Mouse" [M. Green, ed.], pp. 1–7. Gustav Fischer Verlag, New York, 1981) for mouse genes; and the format of the *Proceedings of the National Academy of Sciences* for other names. Sharp-eyed readers will note that the numbers that identify restriction endonucleases were accidentally italicized in the final version of most of the papers. (Only the three-letter names actually belong in italics.) We apologize for this error and trust that it will not lead to any ambiguities.

Acknowledgments

This Birth Defects Symposium would have been difficult, if not impossible, to manage without (1) the tireless administrative efforts of Luba Goldin, and her assistant, Kathy Ruth; (2) the cooperation of the entire clerical staff, particularly in the crucial days just before the meeting took place; (3) the technical assistance of Gary Snodgrass, Kathy Hatch, and, particularly, Paul Maskin; and (4) the helpful advice of last year's symposium chairman, Dr. Ann Willey.

Dorothy Fischer was not only responsible for typing the entire text and setting all of the tables and figures, but she also coordinated the production of the manuscript according to a stringent timetable. Our thanks for a job extremely well done. Norma Hatcher and Kathy Hatch are also due thanks for their proofreading and indexing work, respectively. Dr. Michael Lynes's technical proofreading of the immunology papers was greatly appreciated. We also thank Dr. David Axelrod, Commissioner of Health, and Dr. David Carpenter, Director of the Center for Laboratories and Research, for encouragement and support.

INTRODUCTION TO MOLECULAR GENETICS

Lorraine Flaherty

INTRODUCTION

Over the past ten years the field of recombinant DNA research has been expanding rapidly. Superior new techniques are now available to examine genes and their gene products. These techniques not only have aided basic scientific research; they have also been applicable in diagnosis and treatment of diseases.

Because these techniques are relatively new, few researchers have been formally educated in their use. This review is a brief introduction to the field, so that the material presented in the following chapters can be more readily understood. The review will concentrate on the basic approaches used to isolate and characterize the expression and function of a mammalian gene.

LIBRARIES

The first step in isolating a gene is to prepare mammalian DNA from cells or tissue homogenates. Commonly, DNA libraries are made which contain a large number of genes. The particular gene of interest can be isolated from these libraries.

There are two basic types of libraries: a genomic library and a complementary DNA or cDNA library. The genomic library is a collection of small pieces of DNA, contained within self-replicating vectors, which represent the entire mammalian genome. A cDNA library is also a collection of small pieces of DNA, contained within self-replicating vectors, but it represents only that DNA which is transcribed into messenger RNA (mRNA). Obviously the content of the cDNA library is very dependent on the cells used in its construction, since transcriptional processes are variable among cell types.

Genomic Library

To start a genomic library, DNA is prepared which consists of fragments greater than 100 kilobases (kb) (Figure 1). These fragments are broken up into smaller units by partial digestions with restriction enzymes or, less commonly, by mechanical shearing. Restriction enzymes (Type II) are those which recognize a

ISBN 0-12-492220-1

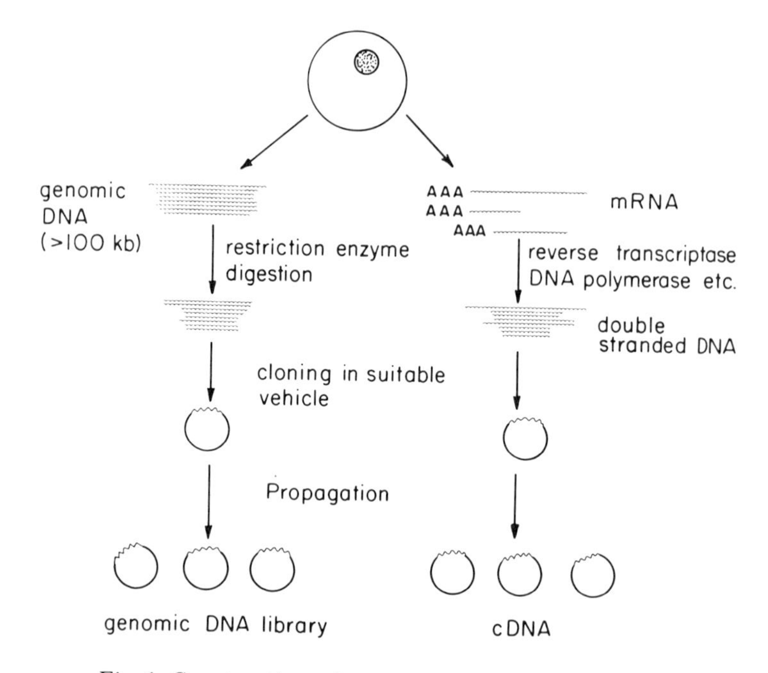

Fig. 1. Construction of a genomic and cDNA library.

unique sequence in the DNA and cleave it at a specific site, either within the recognition sequence or adjacent to it. Some enzymes recognize a 4-base-pair sequence and will cut the DNA more frequently (*e.g. Sau3A*), while others have longer recognition sequences (*e.g. Bam* HI, 6 base pairs). There are several excellent reviews on the nature and use of restriction enzymes.[1-3]

For genomic DNA libraries a random assortment of stretches of DNA is desired. Therefore enzymes which cleave the DNA more frequently are used. If long stretches of DNA are needed for these libraries, only partial digestion conditions are employed, so that not all (or only a few) sites are cleaved. By choosing the right enzyme and varying the exposure time or amount of enzyme, it is possible to obtain DNA fragments in a variety of sizes.

These DNA fragments are inserted into an appropriate vector or vehicle for replication, cloning, and propagation. The choice of vector depends upon the fragment size and particular application of the library.

Three basic types of vectors are commonly used for this purpose: the plasmid, bacteriophage and cosmid (Table 1). All are replicating structures which grow easily in bacteria. They have a number of common properties. For example, they usually contain drug-resistant genes which aid in their isolation and detection. They also contain certain restriction enzyme sites, which aid in the insertion of foreign pieces of DNA. Each vector has size restrictions. pBR322, a commonly used plas-

TABLE 1. Vectors for Molecular Cloning

Vector	Description
PLASMID	
	Extrachromosomal element of closed circular DNA
	Length: 1–200 kb
	Example: pBR322 (4.3 kb)
	Insertion size: 0–10 kb
BACTERIOPHAGE	
	Double-stranded DNA virus
	Length: 40 kb
	Example: *Charon* 4A
	Insertion size: 10–25 kb
COSMID	
	Combination of bacteriophage and plasmid
	Length: 5–10 kb
	Example: pJB8
	Insertion size: 35–45 kb

mid, can only accommodate easily a piece of DNA less than 10 kb long, whereas a bacteriophage, such as *Charon* 4A, can accommodate a fragment of 10–20 kb.

Recently more researchers have been using a new type of vector called a cosmid.[4] Essentially a cosmid is a hybrid between the bacteriophage and plasmid. It contains that part of bacteriophage DNA necessary for insertion and packaging into the head of the bacteriophage (COS sequences), and it contains some of the self-replicating features of the plasmid. For these reasons it can accommodate larger fragments of DNA (approximately 35–45 kb) and is thus useful in studying stretches of DNA and regions adjacent to a known gene. In particular, it facilitates a technique known as chromosomal walking, in which overlapping sequences of several cosmids are used to construct a linear map of a long stretch of DNA.

Once the DNA is inserted into one of these vectors, the vectors are grown in bacteria on agar plates and screened for the desired DNA sequence or gene.

cDNA Library

cDNA libraries are constructed in a similar way except that mRNA instead of DNA is prepared.[5] Since most mRNA's have a poly(A) tail, the mRNA can be separated from other RNA's by passage over an oligo(dt) column. The mRNA is then used to make cDNA by the action of an enzyme, reverse transcriptase, which transcribes RNA into DNA. As in the construction of a genomic library, this DNA is inserted into vectors and propagated.

Colony Hybridization

To select the colony or plaque containing the desired gene, a technique known as colony hybridization is used (Figure 2, ref. 6). A nitrocellulose filter is

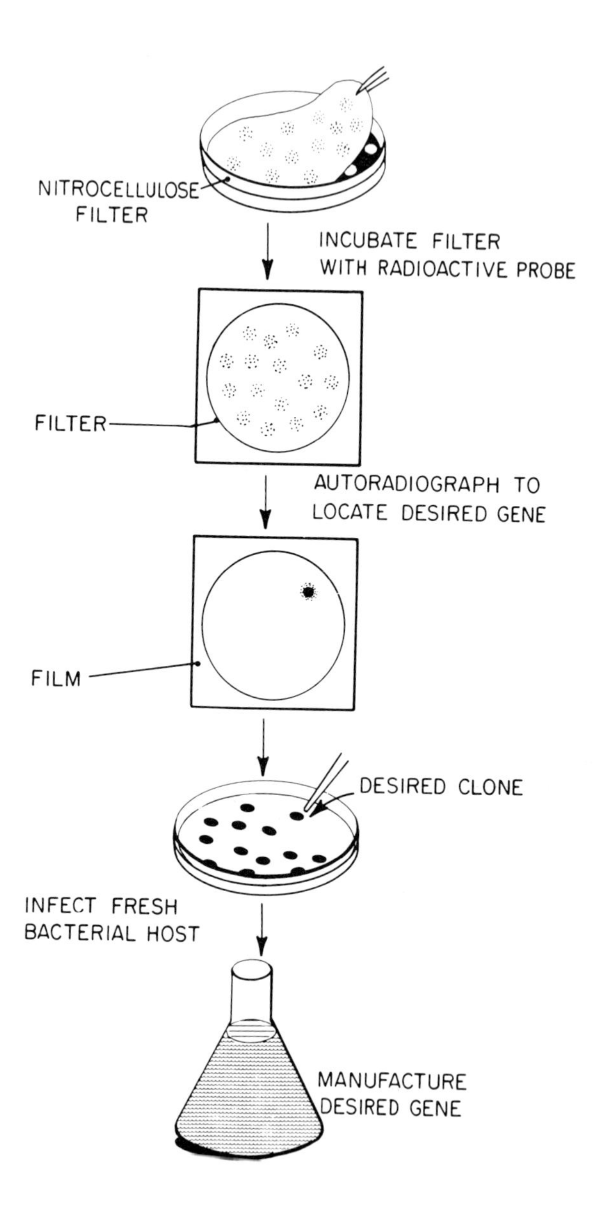

Fig. 2. Colony hybridization (adapted from Leder, ref. 6). Figure reprinted courtesy of *Scientific American.*

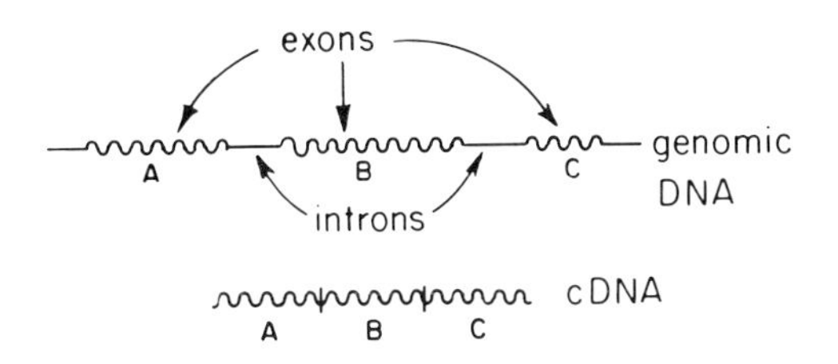

Fig. 3. Comparison of genomic and cDNA.

placed over the bacterial colonies or phage plaques. This filter absorbs small amounts of the DNA from the colonies. The filter is then incubated with a radioactive DNA probe, similar in sequence to part of the desired gene. Any DNA possessing this sequence will hybridize with it. The excess radioactive probe is washed away and the filter exposed to an X-ray film. The position of the exposure on the film will indicate the colony which contains DNA with a hybridizable sequence.

Uses and Precautions

There are vast differences in the applications of these two types of libraries. Most mammalian genes are interrupted by sequences of DNA which are not transcribed into mRNA (Figure 3). The coding regions of a gene are called exons; the intervening and noncoding regions, which are not transcribed into RNA, are called introns. These introns, however, do not exist in the cDNA, since they are eliminated during the formation of the mRNA. Comparisons of the genomic and cDNA of a particular gene can thus be informative about the organization of the genetic material, as well as about its transcription.

In the construction of these libraries it is important to consider whether the entire collection of DNA is accurately represented.[4,5] First, it is necessary to produce a large number of clones, so that all stretches of DNA are represented. Statistically, if the fragments are 20 kb long, the number of clones required to represent the entire genome once is 150,000 ($P = 0.5$).[4] To ensure that all genes are obtained, it would be necessary to screen 3–10 times this number. Second, during the production or propagation of the library, mutation and/or recombination can occur. These are rare events, however, and precautions such as the use of the recombinant-deficient bacteria can ensure more fidelity. Finally, comparison of the restriction enzyme pattern of total genomic DNA (or cDNA) with the DNA obtained from the library can be used to confirm fidelity.

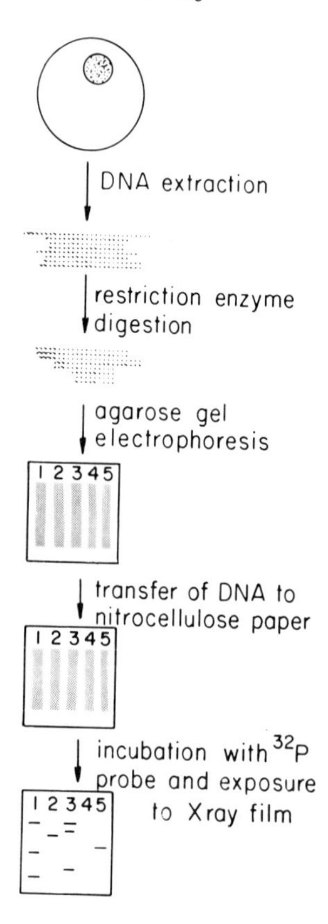

Fig. 4. Southern blot analysis.

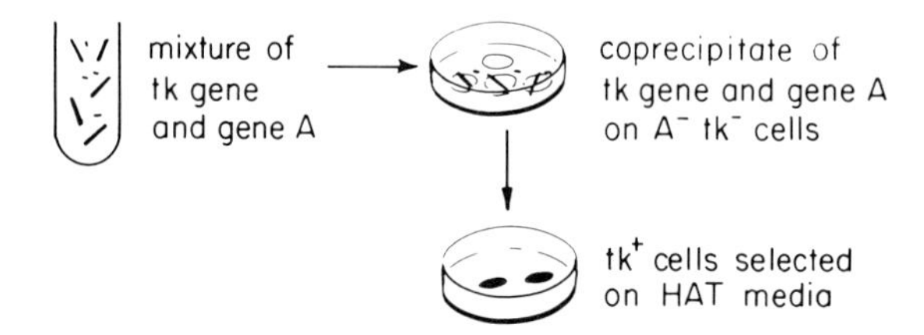

Fig. 5. DNA-mediated gene transfer.

ANALYSIS

Two types of analyses have been very important for genetic and functional studies of genes: Southern blot analysis and gene transfer techniques.

Southern Blot Analysis

Southern blots were named after the scientist who first described them, E. Southern.[7] They are often used to determine the polymorphism of a stretch of DNA in or near a particular gene (Figure 4). In the first step, genomic DNA is completely digested with a restriction enzyme. Since the recognition site of the enzyme occurs at different positions along the DNA, this procedure generates an assortment of different-sized fragments. Next the DNA is separated on the basis of size by agarose gel electrophoresis. The DNA appears as a smear of different sizes. It is then transferred to nitrocellulose and incubated with an appropriate radioactive DNA probe. Any DNA which contains a homologous sequence will be indicated after exposure to an X-ray film.

A) GENE TRANSFER TO EMBRYOS

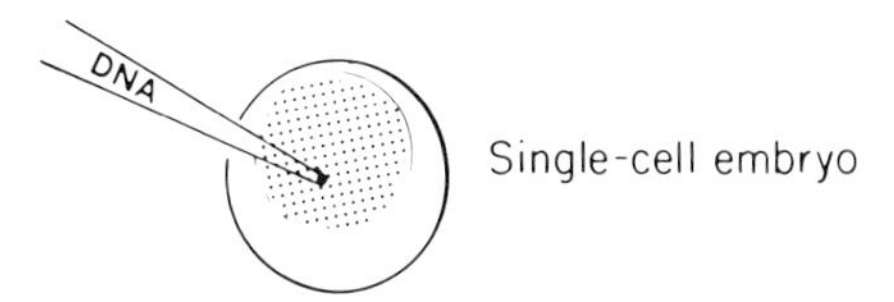

B) FUSION OF LIPSOMES WITH CELLS

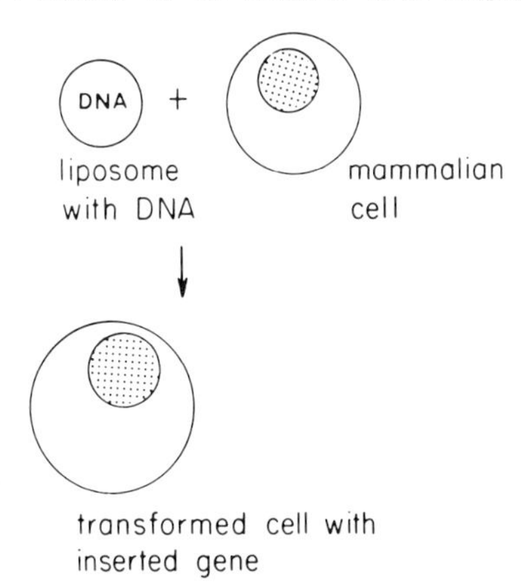

Fig. 6. Other methods of introducing foreign DNA into a mammalian cell.

In this manner the polymorphism of a DNA segment can be studied. For example, one individual may have pattern one and another pattern four (Figure 4). Multigene families can also be detected by this analysis.

Gene Transfer

Various techniques have been devised to introduce isolated genes into mammalian cells to study the genes' function.[8] Perhaps the most popular is DNA-mediated gene transfer (Figure 5) with the use of calcium phosphate. DNA's are prepared from the isolated gene and from the thymidine kinase (*tk*) gene. The tk enzyme is essential for growth of cells on a selective medium containing HAT (hypozanthine, aminopterin, and thymidine). These two DNA's are then coprecipitated and placed on a culture of mammalian cells (for example, mouse L cells). Certain of these cells take up the DNA and are selected by growth in HAT medium. Once the gene has been transferred, its expression can be studied. This technique has proven successful for several genes including major histocompatibility genes of the mouse and man (see Seidman and Orr, this volume).

Other methods are also becoming available and these have certain advantages. One method, microinjection of embryos (Figure 6), offers the advantage that a gene can be studied *in vivo*.

CONCLUSIONS

This new technology offers an opportunity to isolate any mammalian gene and study its function. Recombinant techniques are currently being used for diagnosis of hemoglobinopathies and other genetic diseases. The future applications are endless. Conceivably techniques will soon be available, not only to diagnose most genetic diseases in a superior manner to those previously used, but also to correct them by manipulation of the cellular genetic material.

REFERENCES

1. Maniatis T, Fritsch EF, Sambrook J. Molecular Cloning. A Laboratory Manual. Long Island:Cold Spring Harbor Laboratory, 1982:98-106.
2. Malcolm ADB. The use of restriction enzymes in genetic engineering. *In*: Williamson R, ed. Genetic Engineering. Vol. 2. New York:Academic Press, 1981:130-73.
3. Roberts RJ. Directory of restriction endonucleases. *In*: Methods in Enzymology. Vol. 68. New York:Academic Press, 1979:27-41.
4. Dahl HH, Flavell RA, Grosveld FG. The use of genomic libraries for the isolation and study of eukaryotic genes. *In*: Genetic Engineering Vol. 2. New York:Academic Press, 1981:50-127.
5. Williams JG. The preparation and screening of a cDNA clone bank. *In*: Genetic Engineering Vol. 1. New York:Academic Press, 1982:2-59.
6. Leder P. The Genetics of Antibody Diversity. *Sci Am* 1982; 246:102-15.
7. Southern E. Detection of specific sequences among DNA fragments separated by gel electrophoresis. *J Mol Biol* 1975; 98:503-17.
8. Scangos G, Ruddle FH. Mechanisms and applications of DNA-mediated gene transfer in mammalian cells; a review. *Gene* 1981; 14:1-10.

THE MOLECULAR BASIS OF FAMILIAL GROWTH HORMONE DEFICIENCY

John A. Phillips III

INTRODUCTION

Human growth hormone (GH) is a 191-amino acid polypeptide hormone which is secreted by a subpopulation of the acidophilic cells (somatotropes) of the anterior pituitary. Release of GH into the peripheral circulation is enhanced by growth hormone releasing factor which is produced by the hypothalamus. A deficiency of GH production causes metabolic alterations and growth failure.[1] While in most cases GH deficiency is idiopathic, there are six known single gene disorders that are associated with a deficiency or absence of GH (see Table 1).[2-5] These disorders differ in their phenotype (degree of GH deficiency as well as presence of other hormonal deficiences) and their modes of inheritance. This heterogeneity could be due to the fact that these different disorders affect different steps in GH synthesis and release. For example, one genetic defect could alter the growth hormone releasing factor, another the cells of the hypothalamus necessary for production of this factor, another differentiation of function of the somatotropes of the anterior pituitary, and finally the *GH* gene itself could be abnormal. The possible diversity in defects causing GH deficiency suggested by Table 1 can be reduced to three simple mechanisms (Table 2). First, GH deficiency could be due to a simple deletion of the *GH* structural gene. Second, mutations within or near the *GH* genes could cause an altered *GH* gene product with decreased immunological and biological potency or a deficient amount of structurally normal GH peptide. Third, GH deficiency could be associated with normal *GH* genes that are rendered nonfunctional by mutations at distant loci that control factors important in GH production.[2,6] Before giving examples of two of these types of defects, I will briefly review some of what is known about the *GH* gene cluster.

The GH and human chorionic somatomammotropin (CS) are closely related polypeptide hormones that have 92% homology between their mRNA coding sequences.[7,8] Following digestion with *Eco*RI the *GH* and *CS* genes are contained in 2.6- and 2.9- kilobase (kb) fragments of genomic DNA, respectively, while a third uncharacterized fragment (9.5-kb) contains additional homologous

ISBN 0-12-492220-1

TABLE 1. Genetic Disorders with hGH Deficiency

IGHD	Mode	Endogenous hGH
1A	Autosomal Recessive	Absent
1B	Autosomal Recessive	Decreased
II	Autosomal Dominant	Decreased
III	X-linked	Decreased
PIT Dwarfism		hGH, TSH, ACTH, LH, FSH
I	Autosomal Recessive	Decreased
II	X-linked	Decreased

TABLE 2. Molecular Basis of GH Deficiency

Deletion GH gene	GH
Mutation GH gene	GH*
Another gene ⤳	GH

or "GH-like" (*GH-L*) sequences.[8,9] Studies of the nucleotide sequences of isolated genomic clones suggest there are at least two *GH* and two *CS* loci.[8,10,11] Restriction analyses of genomic DNA indicate that a minimum of five loci (two *GH*, two *CS* and one *GH-L*) exist.[5] In the case of the GH genes, one locus (*GH-N*) encodes the known protein sequence, while the other locus (*GH-V*) encodes a protein which differs by 14 amino acids.[12,13] Although all of these components of the GH gene cluster are located on the long arm of chromosome 17, (specifically, bands q22→24), the arrangement of the components relative to each other is not known.[9,14,15]

To determine the mechanism by which the genetic disorders shown in Table 1 caused hGH deficiency, restriction analysis of DNA rather than sequencing was used. The basis of IGHD-1A was found to be deletion of the *GH-N* gene but in the remaining disorders the *GH-N* genes were not deleted. In IGHD-1B as well as pituitary dwarfism I (Table 1), the cause of GH deficiency is shown to be other genes not linked to the *GH-N* locus (Table 2).

RESULTS

IGHD 1A: When Dr. Ruth Illig described this disorder, she felt it was distinguished by several clinical features. These included familial incidence, early and severe dwarfism, typical appearance, good initial response to exogenous GH followed by development of GH antibodies in sufficient titer to inhibit the therapeutic growth effect and cause growth arrest.[16-18] These observations led to her

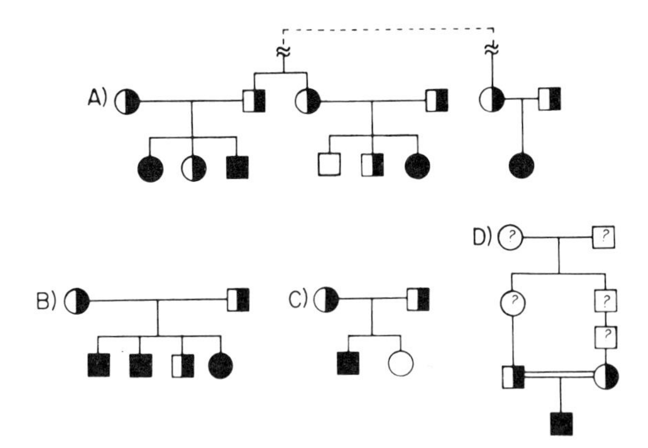

Fig. 1. Pedigrees of families, each of which has one or more individuals affected with IGHD 1A (●, ■). A - Switzerland, families I-III, B - Argentina, C - Austria and D - Japan.

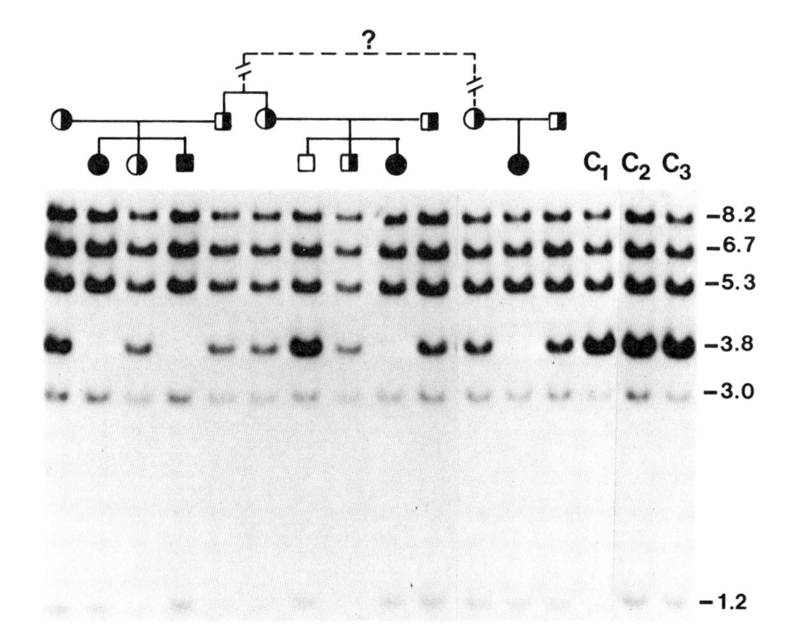

Fig. 2. Autoradiogram patterns of DNA's from families I-III (see Figure 1A) and three controls (C_{1-3}) after digestion with *BamHI* restriction endonuclease. Fragmant sizes in kb are on right.[4]

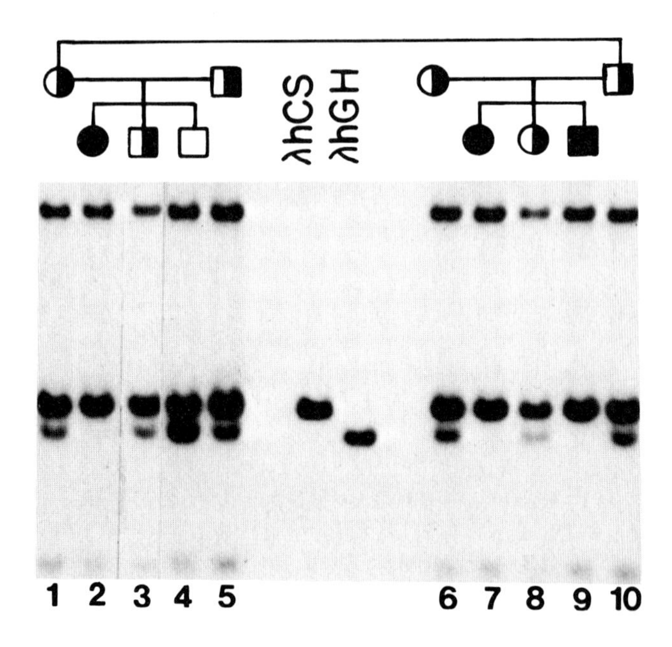

Fig. 3. Autoradiogram patterns of DNA's from families I and II (see Figure 1A) and genomic clones containing either the *CS* gene (λCS) or the *GH-N* gene (λGH) after digestion with *Eco*RI and *BamHI* restriction endonuclease.[4]

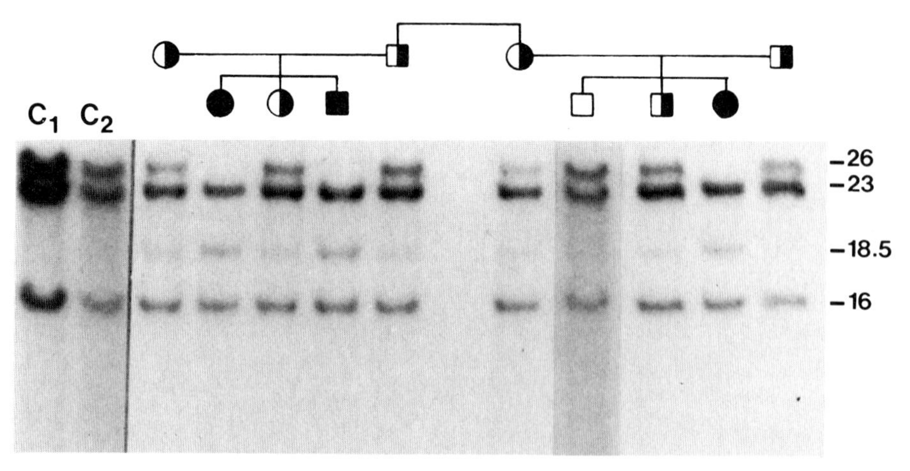

Fig. 4. Autoradiogram patterns of DNA's from families I and II (see Figure 1A) and two controls (C_{1-2}) following digestion with *HindIII* restriction endonuclease. Fragment sizes in kb are on the right.[4]

TABLE 3. Size and Content of Genomic DNA Fragments Containing Components of *GH* Gene Cluster

Size (kb)	Gene(s) contained
26	*GH-N* & *GH-L*
23	*GH-V* & *CS*
16	*CS*

hypothesis that individuals with IGHD 1A had a prenatal deficiency of GH that caused a lack of immune tolerance to exogenous GH.

The pedigrees of three Swiss families, each of which has one or more individuals affected with IGHD 1A, is shown in Figure 1a. Nuclear DNA's from individuals in these families were studied by restriction endonuclease analysis. The autoradiographic patterns of these DNA's were compared to those of three controls ($C_{1\text{-}3}$) after digestion with *BamHI* and hybridization to ^{32}P-labeled *GH* cDNA sequences (Figure 2). Patterns seen after *Eco*RI and *BamHI* digestion of DNA's from these families and from genomic clones containing either the *CS* (λCS) gene or the *GH-N* (λGH) gene indicated each affected was homozygous for a deletion that includes *GH-N* but not *GH-V* sequences (Figure 3). To determine the size of the deletion *HindIII* digestions were used which in normals yield 26-, 23- and 16-kb fragments which contain components of the *GH* gene cluster (Table 3). We assumed the minimum size of the *GH-N* deletion would be reflected by a reduction in the size of the 26-kb fragment. Autoradiogram patterns following Hind III digestion indicated the deletion included a minimum of 7.5-kb of DNA because the 26-kb fragment was replaced by an 18.5-kb fragment (26-18.5 = 7.5-kb) (Figure 4). Restriction patterns of DNA's from all family members agreed with an autosomal recessive mode of inheritance and correlated with the clinical phenotype. Furthermore, independent assortment of the *GH-N* and *-V* genes suggested these genes were non-allelic. Thus, in these families, IGHD 1A is caused by deletion of the *GH-N* genes and hGH deficiency occurs despite the presence of the *GH-V* genes.[4]

Subsequently, samples from additional families in Argentina, Austria and Japan have been studied (Figure 1b-d).[19] Affected individuals in each of these additional families were also homozygous for deletions of their *GH-N* genes. Using *MspI* and *BglII*, we found four common polymorphic restriction sites that occur within or adjacent to components of the *GH* gene clusters in these families as well as normal controls.[5] Interestingly, affected individuals from families, three of the four families shown in Figure 1, had different distributions or patterns of polymorphic *BlgII* or *MspI* restriction sites linked to the remaining components of their *GH* gene clusters (Table 4).

IGHD 1B: The pedigrees of 12 non-related families with IGHD 1B whose DNA's were studied are shown in Figure 5. In each family two sibs were affected with IGHD 1B and the parents were of normal height. Our criteria for IGHD 1B included:

TABLE 4. GH-N Deletions and Flanking Polymorphic Sites

Origin	*BglII*		*MspI*	
	A	B	A	B
Argentina	+	–	–	–
Switzerland	+	–	–	–
Japan	–	+	+	+
Austria	+	+	+	+

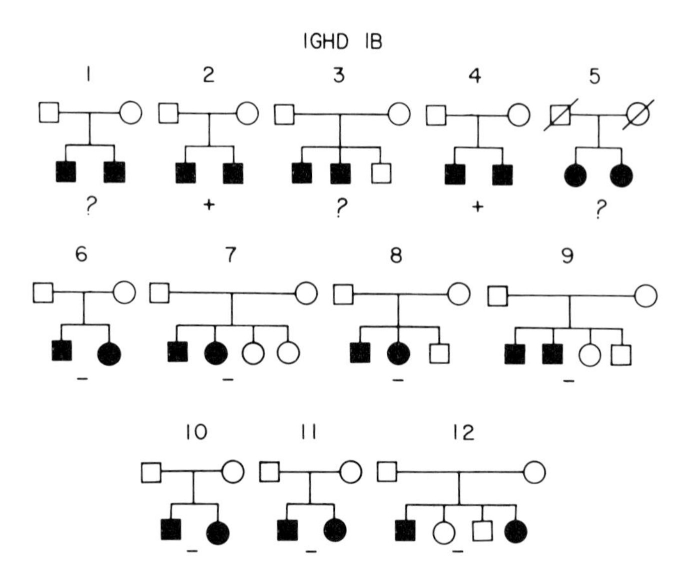

Fig. 5. Pedigrees of twelve families each of which has two sibs affected with IGHD 1B (●, ■). Symbols below pedigree indicate whether affected sib pairs have concordant (+), possibly concordant (?), or discordant (–) polymorphic restriction site patterns or haplotypes.

a) no demonstrable anatomic cause for IGHD deficiency,
b) stature greater than 2 S.D. below mean for age and sex,
c) significantly delayed bone age,
d) peak GH levels less than 7 ng/ml after standard pharmacologic stimulation,
e) deficient growth velocity that responded to exogenous GH,
f) normal thyroid function,
g) spontaneous pubertal changes at an appropriate bone age, and
h) no known immunodeficiency or history of recurrent infection in cases from pedigrees having only affected males.[5]

The *MspI* and *BglII* polymorphic restriction sites mentioned previously and an additional site (*HincII*) were used as markers in linkage analysis of DNA's from family members. Among the 12 pairs of sibs affected with IGHD 1B shown in Figure 5, those in families 6–12 were discordant for the polymorphic restriction sites inherited from their parents (Figure 6). Since *HincII* and *MspI* sites are very close to *GH-N* (116 bp and ⩽17 kb, respectively) we conclude that meiotic recombination is an unlikely cause for this discordance. A more likely explanation for the discordance between the disease state and the polymorphic restriction sites inherited by sib pairs in families 6–12 is that the mutation causing the IGHD 1B phenotype and the *GH-N* structural gene assort independently. This suggests that the mutation involved in IGHD 1B affects a non-linked gene (possibly one responsible for development or regulation of somatotropic cell function) rather than the hGH-N structural gene.[5]

In the remaining five families (families 1–5, Figure 5) the affected sib pairs were definitely concordant only in families 2 and 4. In families 1, 3, and 5 discordance remains possible because the markers are not sufficiently informative to prove concordance (Figure 7). Among the 12 affected sib pairs, two were definitely concordant, three were possibly concordant, and seven were definitely discordant for polymorphic restriction site patterns or haplotypes. These results suggest that most cases of IGHD 1B are due to alterations not linked to the *GH-N* gene. The number of concordant sib pairs observed may reflect the 25% occurrence of concordance expected for independently assorting characteristics or they may reflect a minority of cases that are due to mutations which alter the *GH-N* gene.[5]

Hypopituitary Dwarfism: There are at least two genetic types of pituitary dwarfism, one having an autosomal recessive (AR) and the other an X-linked recessive mode of inheritance (Table 1).[20] When we studied DNA samples from a Hutterite family with the AR type, the polymorphic restriction sites and hence the *GH* alleles inherited did not correlate with the disease phenotype (R. McArthur and J.A. Phillips, unpublished data). This suggests the mutation involved is distant

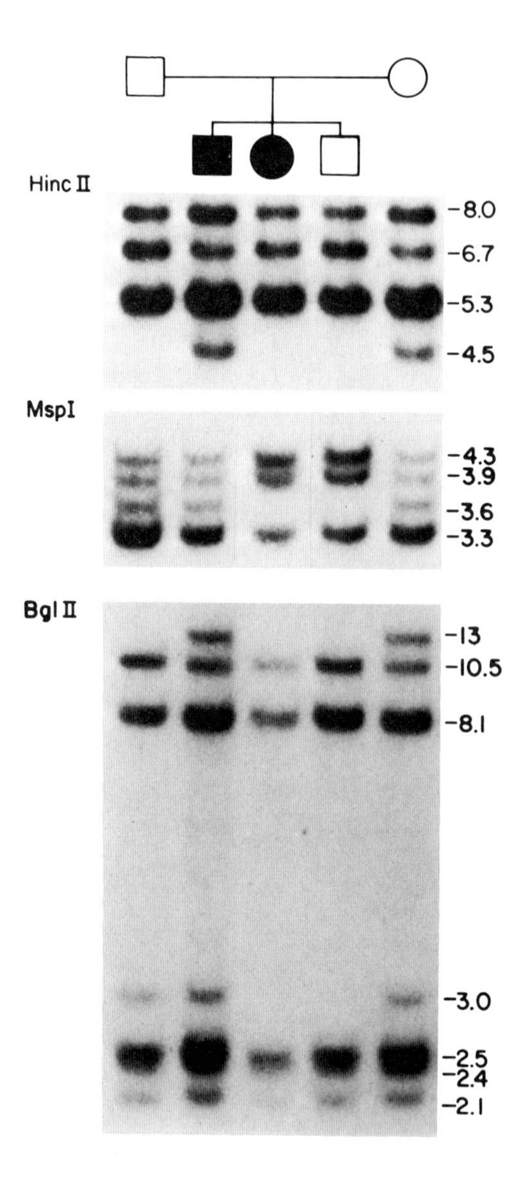

Fig. 6. Autoradiogram patterns of DNA's from family 8 (Figure 5) after digestion with *HincII* (top), *MspI* (middle), or *BlgII* (bottom) restriction endonuclease. Fragment sizes in kb are on the right (*MspI* fragments $\leqslant 3.3$ kb are not shown).[5]

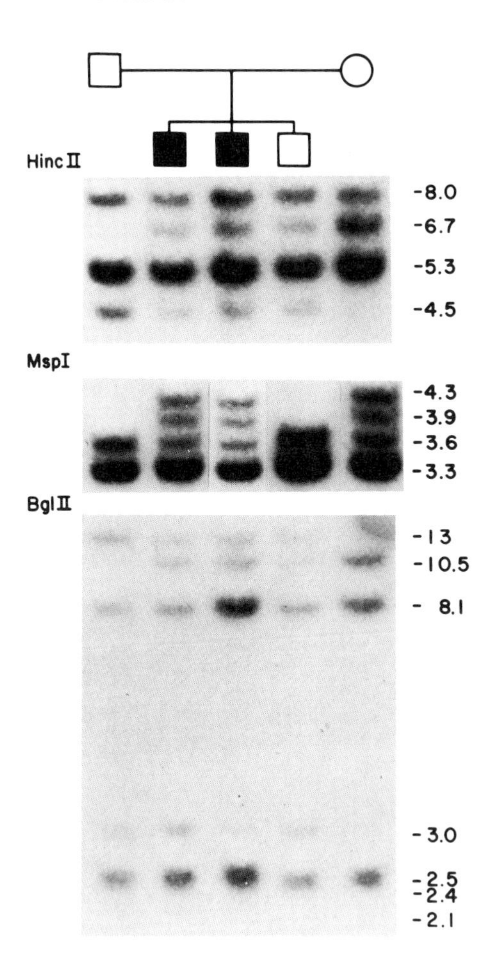

Fig. 7. Autoradiogram patterns of DNA's from family 3 (Figure 5) after digestion with *HincII* (top), *MspI* (middle), or *BglII* (bottom). Fragment sizes in kb are on the right. (*MspI* fragments ⩽3.3 kb are not shown).[5]

to the *GH-N* structural gene. In the case of the X-linked type of hypopituitary dwarfism the phenotype is distinct from IGHD III suggesting there are at least two loci on the X-chromosome that control factors important in GH production.

DISCUSSION

The molecular basis of IGHD 1A was found to be an example of the simple gene deletion shown in Table 2. Since this deletion precludes the production of any GH-N peptide, the propensity of the affected homozygotes to form anti-hGH antibodies is clarified. Their absence of GH-N peptide probably makes them immunologically intolerant of exogenous hGH in analogy to other mutations which lack cross-reacting material (CRM). For some families this poor response to treatment might be sufficient indication for prenatal diagnosis which could be done by DNA analysis of fetal amniotic fluid cells. The distributions of polymorphic restriction sites which are linked to the non-deleted components of the *GH* gene clusters of homozygotes from each family (Table 4) suggest three of the four deletions are associated with different haplotypes. Since different mutations causing β-thalassemia are often associated with specific polymorphic restriction site haplotypes,[21] the deletions may differ in most of our families. If this is true, it implies that recurrent *GH-N* deletions have occurred within the *GH* cluster or that extensive recombination has followed an original deletion. An obvious mechanism of recurrent deletions of the same size would be unequal crossing-over after mispairing of duplicated or analogous sequences such as has been documented for α-globin genes.[22] Further data on the organization of the *GH* gene cluster will be needed to establish the nature and location of such sites of recurrent recombination.

Our experience with IGHD 1B suggests a need for caution in the use of restriction site polymorphisms for diagnosis and prediction of inherited diseases. Such use should be predicated upon study of a sufficient number of families to establish linkage and homogeneity of the disease. For example, conclusions based upon definite or possible concordance between restriction patterns and disease in families 1-5, Figure 5, would have led to erroneous prediction of the absence of disease in the second affected sibs in families 6-12.[5]

Analysis of polymorphic restriction patterns provides a powerful approach for establishing the absence of linkage between a particular disease and a specific DNA sequence within a single family. Using a relatively rapid and inexpensive technique, we determined that *GH-N* gene sequencing would not succeed in establishing the molecular basis of either IGHD 1B or the autosomal recessive type of pituitary dwarfism in the families studied (Table 1, Figures 5,6). In these families any sequence changes that might be found would probably be irrelevant to their disease. Other disorders involving growth for which this approach could be used to select individuals whose DNA would be appropriate for cloning and sequencing of the *GH-N* gene include "biologically inactive" GH[23] and IGHD II[2] (Table 1).

SUMMARY

There are six distinct Mendelian disorders that have associated GH deficiency. One of these (IGHD 1A) is caused by deletion of the *GH* structural gene (*GH-N*). In the two remaining disorders examined, the *GH* gene cluster is grossly intact and for these polymorphic restriction sites within or adjacent to components of the *GH* cluster were used as "markers" in linkage analysis. In IGHD 1B and the autosomal recessive type of pituitary dwarfism affected sibs were discordant for the "markers", hence, the *GH-N* alleles inherited. These results demonstrate the usefulness of linkage analysis in confirming that genetic disorders with GH deficiency may be due to alterations in the *GH* gene before DNA sequencing is done.

ACKNOWLEDGMENTS

The author thanks P. H. Seeburg and J. S. Parks for assistance and advice and J. Bailey, R. Blizzard, H. Frisch, T. Kelly, L. Kirby, P. Malvaux, Y. Nishi, M. Rivarola, C. Sultan, J. Tze, and M. Zachmann for providing clinical samples. I also thank Sandy Muscelli for her expert assistance in preparing the manuscript.

This work was supported in part by the National Institutes of Health grants AM00423, AM28246 and RCDA AM00958.

Figures 2, 3, and 4 appear courtesy of *Proc Natl Acad Sci USA.*[4] Figures 6 and 7 appear courtesy of *J Clin Invest.*[5]

REFERENCES

1. Underwood LE, Van Wyk JJ. Hormones in normal and aberrant growth. *In:* Williams RH, ed. Textbook of Endocrinology. Philadelphia: WB Saunders Co. 1981: 1149.
2. Rimoin DL. Hereditary forms of growth hormone deficiency and resistance. *Birth Defects* 1976; 12:15.
3. Fleisher TA, *et al.* X-linked hypogammaglobulinemia and isolated growth hormone deficiency. *New Engl J Med* 1980; 302:1429.
4. Phillips III JA, Hjelle BL, Seeburg PH, Zachmann M. Molecular basis for familial isolated growth hormone deficiency. *Proc Natl Acad Sci* 1981; 78:6372.
5. Phillips III JA, *et al.* Genetic analysis of familial isolated growth hormone deficiency type I. *J Clin Invest* 1982; 70:489.
6. Rimoin DL, Schechter JE. Histological and ultrastructural studies in isolated growth hormone deficiency. *J Clin Endocrinol* 1973; 37:725.
7. Niall HD, Hogan ML, Sauer R, Rosenblum IY, Greenwood FC. Sequence of pituitary and placental lactogenic and growth hormones: Evolution from a primordial peptide by gene reduplication. *Proc Natl Acad Sci* 1971; 68:866.
8. Fiddes JC, Seeburg PH, DeNoto FM, DeNoto FM, Hallewell RA, Baxter JD, Goodman HM. Structure of genes for human growth hormone and chorionic somatomammotropin. *Proc Natl Acad Sci* 1979; 76:4294.
9. Owerbach D, Rutter WJ, Martial JA, Baxter JD, Shows TB. Genes for growth hormone, chorionic somatomammotropin, and growth hormone-like gene on chromosome 17 in humans. *Science* 1980; 209:289.
10. Goodman HM, *et al.* Structure and evolution of growth hormone related genes. *In*: Scott WA, Werner R, Joseph DR, Schultz J, eds. Mobilization and Reassembly of Genetic Information. New York:Academic Press, 1980; 17:155.
11. Seeburg PH. Structure and regulation of pituitary hormone genes. *In:* Beers RF, Bassett EG, eds. Polypeptide Hormones. New York:Raven Press, 1980:19.

12. Pavlakis GN, Hizuka P, Gorden P, Seeburg PH, Hamer DH. Expression of two human growth hormone genes in monkey cells infected by Simian virus 40 recombinants. *Proc Natl Acad Sci* 1981; 78:7398.
13. Seeburg PH. The human growth hormone gene family: Nucleotide sequences show recent divergence and predict a new polypeptide hormone. *DNA* 1982; 1:239.
14. George DL, Phillips III JA, Francke U, Seeburg PH. The genes for growth hormone and chorionic somatomammotropin are on the long arm of human chromosome 17 in region q21→qter. *Hum Genet* 1981; 57:138.
15. Harper ME, Barrera-Saldana HA, Saunders GF. Chromosomal localization of the human placental lactogen-growth hormone gene cluster to 17 q22-24. *Am J Hum Genet* 1982; 34:227.
16. Illig R. Growth hormone antibodies in patients treated with different preparations of human growth hormone (HGH). *J Clin Endocrinol Metab* 1982; 31:679.
17. Illig R, Prader A, Ferrandez A, Zachmann M. Hereditary prenatal growth hormone deficiency with increased tendency to growth hormone antibody formation. *In:*Kracht J ed. Endokrinologie der Entwicklung und Reifung. Berlin:Springer, 1970:246.
18. Illig R, Prader A, Ferrandez A, Zachmann M. Hereditary prenatal growth hormone deficiency with increased tendency to growth hormone antibody formation ("A-type" isolated growth hormone deficiency). *Acta Pediatr Scand Suppl* 1971; 60:607.
19. Rivarola MA, Phillips III JA, Migeon CJ, Hjelle BJ. Phenotypic heterogeneity in familial isolated growth hormone deficiency (IGHD) type A. *Pediatric Res* 1982; 16:143A.
20. McKusick VA. Mendelian Inheritance in Man. Baltimore:Johns Hopkins University Press, 1978:288,639.
21. Orkin SH, *et al.* β-thalassemia mutations, β-globin gene polymorphisms, and their linkage with DNA polymorphisms in the human β-globin gene cluster.*Nature* 1982; 296:627-31.
22. Philliaps III JA, *et al.* Unequal crossing-over: A common basis of single alpha-globin genes in Asians and American Blacks with hemoglobin-H disease. *Blood* 1980; 55:1066.
23. Kowarski AA, Schneider J, Ben-Galim E, Weldon VV, Daughaday WH. Growth failure wtih normal serum RIA–GH and low somatomedin activity: Somatomedin restoration and growth acceleration after exogenous GH. *J Clin Endorinol Metab* 1978; 47:461.

MAPPING OF RESTRICTION FRAGMENT LENGTH POLYMORPHISMS (RFLP LOCI) TO HUMAN CHROMOSOMES IN SOMATIC CELL HYBRIDS

Uta Francke
Bérengère de Martinville

Restriction fragment length polymorphisms (RFLP) reflect the heterogeneity in the length of DNA fragments produced by cutting high molecular weight genomic DNA with specific restriction endonucleases. The polymorphisms are detected by hybridization of labelled single copy probes to DNA fragments on nitrocellulose filters (Southern blot analysis). RFLP can be caused either by base substitutions in the recognition sequence of a specific restriction enzyme, which abolishes that cleavage site, or by insertion or deletion of stretches of DNA within the fragment length. RFLP occur in the normal population. They have been detected in the vicinity of all genetic loci for which cloned DNA probes are available.[1-3] Therefore, it is assumed that they are distributed widely and randomly throughout the human genome. RFLP can be detected using single copy sequences isolated from random genomic or cDNA recombinant gene libraries. RFLP defined by such probes represent arbitrary or anonymous loci not associated with any specific known gene or gene product. RFLP are inherited as Mendelian co-dominant traits.[4] Since their detection does not rely on gene expression they can be studied in any human tissue that contains DNA, *e.g.* white blood cells, fibroblasts, amniotic fluid cells, or any other cultured cells. Tissue samples, such as hair roots or trophoblastic villi, could also be employed.

RFLP have been hailed as a new class of genetic markers that will greatly facilitate the development of a complete genetic map of the human genome.[5] A systematic search for single copy sequences that define specific RFLP loci has begun in several laboratories and close to 50 have already been discovered.[6] At the same time, DNA samples from large families with Mendelian autosomal dominant or X-linked disorders are being collected for linkage studies. Since for most autosomal dominant disorders the biochemical defects are as yet unknown, linkage of the disease genes such as those for Huntington Disease or neurofibromatosis to polymorphic DNA loci could greatly facilitate genetic counseling. The study of

ISBN 0-12-492220-1

RFLP markers in persons at risk would lead to the detection of gene carriers before the onset of symptoms, and may allow one to determine whether a fetus has inherited the disease gene. For family linkage studies and for the application of linkage information in clinical genetics it is not absolutely necessary to know the physical chromosomal location of the respective loci. However, this information can be useful in a number of ways.

REASONS FOR MAPPING RFLP

Our laboratory has been engaged in the mapping of RFLP to specific chromosome regions for a number of reasons.

1. Such an assignment helps in selecting other genetic markers for studying linkage to the RFLP in families. Markers located on the same chromosome will be suitable, while those assigned to other chromosomes can be excluded. This point has been illustrated by our assignment of the first highly polymorphic random DNA locus to the distal half of the long arm of chromosome 14.[7] This physical localization suggested that the locus could be within measurable distance of two other highly polymorphic loci in man: *PI* (a_1-antitrypsin) and *GM* (immunoglobulin heavy chains). Further studies by others, based on our assignment, have in fact shown that the insertion/deletion polymorphism first described by Wyman and White,[4] now called *D14S1*,[7] maps to the most distal band 14q32 and is closely linked but separable from the immunoglobulin heavy chain gene cluster.[8]

2. Physical mapping of RFLP loci will help to distinguish allelic from non-allelic restriction fragments. Sequences on different chromosomes may have enough homology to hybridize with the same single copy probe.[9]

3. The physical mapping of RFLP loci will assure their even distribution in the construction of a complete linkage map. Results so far indicate that the random polymorphic DNA loci that have been mapped are in fact distributed among different chromosomes and chromosome regions.[6]

4. Eventually the precise mapping of RFLP to individual chromosome bands can be correlated with linkage data obtained from family studies. Thus, it will be possible to establish a relationship between the genetic map based on recombination fractions, the cytological map and the molecular map. At this time the relationship between map units in centimorgans, microns on metaphase chromosomes and kilobases of DNA can only be estimated in general terms. Future studies may indicate that the relationship between these parameters differs for different regions of chromosomes.

5. Lastly, RFLP loci that are mapped to specific chromosome regions or individual bands will provide a great boon to clinical cytogeneticists in the identification and characterization of structurally abnormal chromosomes. For example, in individuals with heterozygous partial deletions or partial triplications of chromosome regions one can look for the presence or absence of specific restriction fragments. Furthermore, dosage studies have shown that the intensity of hybridization signals on Southern blots can be related to the number of copies of the respective restriction fragment.[10] Thus, gene dosage studies in the direct sense of the work, (*i.e.* independent of gene expression) have become possible.

APPROACHES TO MAPPING OF RFLP LOCI

A number of approaches have been devised for the mapping of RFLP loci to specific chromosomes.

1. One can use a probe that originates from a chromosome specific recombinant DNA library. Such libraries are prepared either from sorted chromosomes[10,11] or from rodent x human somatic cell hybrids that contain a single human chromosome.[12] The use of such a probe makes it likely that the RFLP detected with it will map to that specific chromosome. However, this cannot be taken for granted, but has to be confirmed by another approach, because sorted chromosome fractions are not 100% pure and somatic cell hybrids may contain unrecognized material from other human chromosomes.

2. Hybridization of a randomly derived probe to DNA from human chromosomes that have been size-fractionated by flow-sorting, suggests a chromosomal location of the probe in a specific size fraction. The power of this approach lies in the use of human cells with defined chromosome rearrangements that cause a shift of the hybridizing fraction to larger or smaller chromosomes depending on the nature of the translocation.[13] The drawback of this method, which limits its use to only a few technologically well-equipped laboratories, consists of the difficulties in obtaining enough specific chromosome fraction material for Southern blot analyses.

3. Direct *in situ* hybridization of labelled DNA fragments to metaphase chromosomes followed by autoradiography has been well established for localizing repetitive sequences such as ribosomal genes or histone genes.[14] For the localization of unique or low copy number sequences, this approach has to rely on statistical analysis of silver grain distributions over a large number of metaphase cells.[15] Few primary gene/chromosome assignments

have been made by this approach, but it has been useful in the more precise regionalization of DNA loci assigned to specific chromosomes by other means.

4. Our approach, as described in this paper, has been Southern blot analysis of DNA extracted from somatic cell hybrid panels that contain partial human chromosome complements as well as known rearranged human chromosomes.[7] Large numbers of such hybrids are available. Hybrids made with any human tissue can be used without regard to transcriptional activity of the sequences homologous to the hybridization probe. No differences in restriction fragment patterns are expected as long as one uses restriction enzymes that do not distinguish between methylated and unmethylated cytosine residues.[16]

The use of somatic cell hybrids allows one to assign a single copy sequence to a single chromosome. By choosing the appropriate hybrids it is possible to rule out the presence of any sequences that also hybridize to the probe on any other chromosome. If, on the other hand, there are (partially or completely) homologous sequences on nonhomologous chromosomes, the study of somatic cell hybrids will clarify the situation unambiguously. In particular, digestion of the somatic cell hybrid DNA with an enzyme that reveals a restriction fragment length polymorphism has allowed us to assign specific fragment alleles to one or the other of morphologically distinguishable pairs of homologous chromosomes; distinguishable, for example, by heteromorphisms or rearrangements.[17] In the case of nonallelic fragments, we have been able to assign specific size fragments to different chromosomes, *e.g.* the X and the Y chromosome.[9]

Over the years, we have established hybrid clone panels for intrachromosomal gene mapping derived from 15 human donors with defined chromosome abnormalities , and different rodent cell lines. Human parental cell origins include peripheral blood leukocytes, leukemic cells, skin fibroblasts and fetal liver cells. The panel now available for regional assignments of gene loci involves 53 regions on 17 different chromosomes.[18] The following examples will demonstrate the use of these somatic cell hybrid panels for the mapping and regional localization of RFLP loci.

D14S1

In collaboration with A. Wyman and R. White, then at the University of Worcester, we have determined the chromosomal location of the highly polymorphic insertion-deletion RFLP previously described by them.[4] Human/rodent somatic cell hybrid clones were chosen from six different series of hybrids that were derived from six different human donors and two established Chinese hamster cell lines. The content of the human chromosomes in these hybrids at the time of DNA extraction was determined by karyotyping 12 to 30 trypsin-Giemsa banded hybrid metaphase spreads. Thus, the average copy number of each human chromosome was established in each clone. The distribution of the human chromosome content in some of the hybrids is shown diagrammatically in Figure 1. The solid boxes indicate human chromosomes present that were apparently intact. The hatched boxes indicate that only a defined part of the chromosome was present. A cross means low copy number: only 10% or fewer of the cells contained the respective chromosome.

DNA from the hybrids was tested by Southern blot analysis for the presence of sequences homologous to the single copy human insert in recombinant plasmid pAW101.[4] The presence of homologous sequences in the hybrids correlated exclusively with the presence of human chromosome 14. Counting only the intact chromosomes in these and other hybrids studied, we determined the fraction of discordant hybrid clones for each of the other chromosomes. It ranged between .24 and .8 and was greater than .44 for all but three chromosomes.[7] In using somatic cell hybrids for gene assignments it is essential not only to show concordancy between the gene to be mapped and a single human chromosome, but also to rule out each other chromosome as a possible site of the gene. This becomes especially important when the "gene" to be mapped is a DNA sequence homologous to a single copy probe. As is being discovered now, many DNA sequences are not truly unique but are members of a gene family, some of which have dispersed copies on other chromosomes.[19]

In the case of *D14S1* the polymorphism was particularly useful since five of the six human hybrid cell donors were heterozygous, producing two equal intensity *EcoRI* fragments on Southern transfers.[7] Three hybrids contained chromosome 14 at a frequency of greater than one per cell and exhibited both restriction fragment alleles. The average copy number of greater than one for chromosome 14 must reflect the presence of both homologues of chromosome pair 14 in these hybrids. The quantitative analysis of the human chromosome content thus provided additional confirmation for the assignment of the RFLP to chromosome 14.

The regional assignment of *D14S1* to the distal half of the long arm of chromosome 14 (region q21→qter) was based on two series of hybrids in which the human donors carried a balanced t(X;14)(p22.2;q21) translocation. The derivative X chromosome carrying region 14q21→qter contained the active HGPRT gene and was selected for in HAT medium (Figure 2) and counter-selected in medium containing guanine analogues.[20] A biochemical marker for this chromosome was the human G6PD expression as determined by electrophoresis. The derivative chromosome 14 had the size of a human 21 but was distinguishable by banding pattern. Human purine nucleoside phosphorylase, the gene which had previously been assigned to the proximal part of 14q,[20] was used as a biochemical marker for this chromosome. Hybrids XIII-5B, XIII-1B and XIII-7A contained

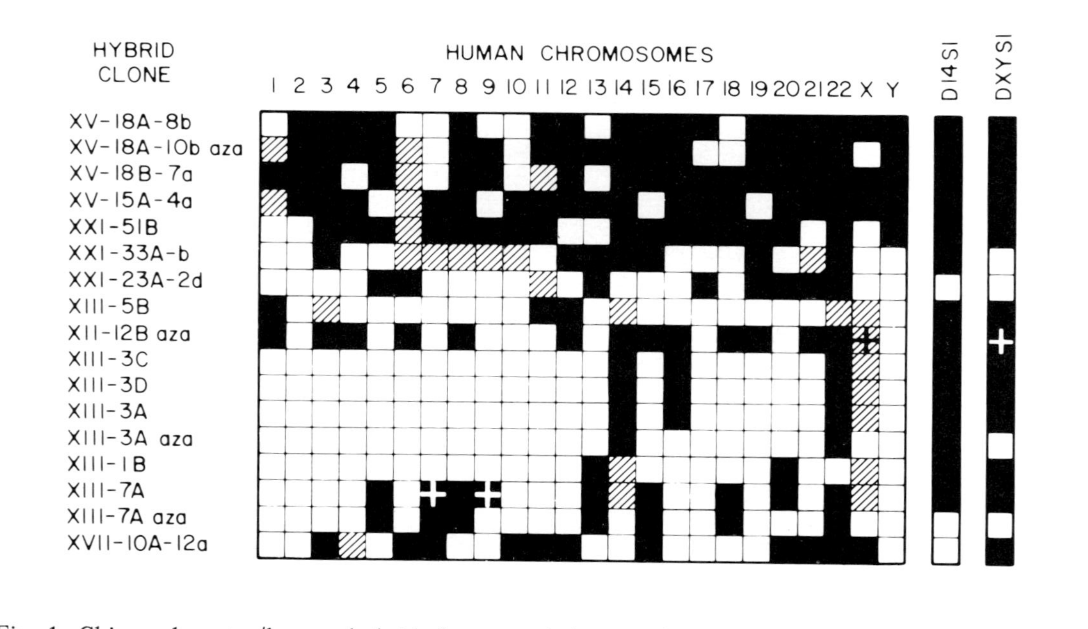

Fig. 1. Chinese hamster/human hybrid clone panel: human chromosome content and hybridization with probes pAW101 (locus *D14S1*) and pDP31 (locus *DXYS1*). Dark boxes indicate presence of intact human chromosome and positive signal on Southern transfers. Hatched boxes indicate presence of only a defined region of the respective chromosome. Crosses indicate low frequency of chromosome (average copy number 0.1 or less), and weak hybridization to probe.

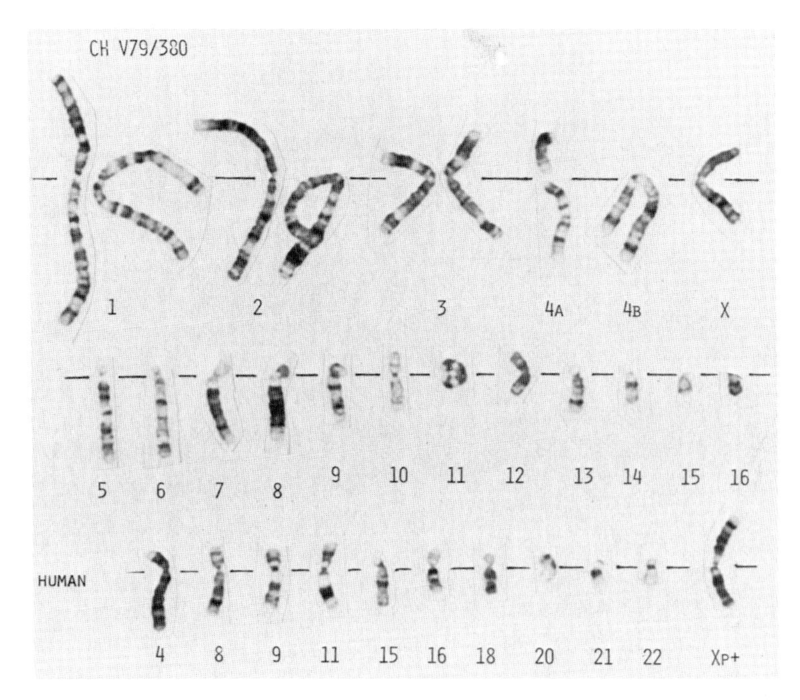

Fig. 2. Trypsin-Giemsa banded karyotype of Chinese hamster/human hybrid of series XII. Human chromosomes are arranged in bottom row. The designation Xp+ refers to the derivative chromosome der(X),t(X;14)(14qter→14q21: :Xp22.2→Xqter). The normal chromosome 14 and the der(14) are not present in this hybrid clone.

the derivative X in the absence of both the derivative 14 and the normal 14 (Figures 1,2). Their DNA hybridized with the pAW101 probe thus assigning *D14S1* to distal 14q.[7]

DXYS1

The same filters of *Eco*RI digested DNA's of this somatic cell hybrid panel (Figure 1) were analyzed by D. Page and A. Wyman in D. Botstein's laboratory at the Massachussetts Institute of Technology (MIT) for hybridization with an apparently single copy fragment of the human insert in recombinant phage λ-rHs4813 from the Maniatis human genomic library. It was interesting to map the sequence because it had been shown to detect a *TaqI* polymorphism in human DNA. A single 4.5kb *Eco*RI fragment was observed in the majority of hybrids. However, the presence of the sequences did not seem to correlate with the presence of a single human chromosome (Figure 1). The lowest discordancy rate was seen with the X chromosome, including those hybrids which had the derivative X chromosome of the t(X;14) translocation with region Xp22.2→Xqter. Two hybrids derived from two different series were positive and did not contain the X chromosome but contained a large number of other human chromosomes (hybrids XV–18A–10b aza and XXI–51B; Figure 1). The results were tested for pairwise correlation assuming sequences homologous to the probe on the X plus on any other of the human chromosomes. The only concordancy seen on the pairwise analysis was for the X and the Y chromosome.[18]

To confirm this assignment we analyzed, in collaboration with D. Page, a number of additional hybrids derived from human male donors. Hybrid DNA was digested with the enzyme *TaqI* which reveals restriction fragments of 14.6kb, 11.8kb and 10.6kb in human DNA. Without exceptions, the results indicated that the 14.6kb allele was carried by the human Y chromosome while the 11.8kb and 10.6 kb alleles were associated with the human X chromosome; in particular with region Xp22.2→Xqter. These findings were in agreement with the results obtained with random human DNA samples: all males carried the 14.6kb fragment, while females had either one or both of the other two fragment sizes.[9] This was the first observation of the presence of unique DNA sequences on both human sex chromosomes. We do not know at the present time the extent of the apparently homologous region nor its precise localization on the X and Y chromosomes. Our hybrid data have excluded the distal one-fourth of the short arm of the X chromosome as a possible site. Initially, this seemed surprising, since it is the distal short arm of the X that pairs with the short arm of the Y in meiosis. On the other hand, if the conserved sequences were located in the regions where the X and Y chromosomes synapse, one would expect to see occasional recombination. This was not observed in the analysis of segregation of the *TaqI* fragment sizes in a 48 member Mormon kindred studied by D. Barker and R. White.[9] This example demonstrates the usefulness of somatic cell hybrids in the assignment of specific restriction fragments to non-homologous chromosomes. According to recommended nomenclature the homologous sequences should be designated *DXYS1*.[6]

The availability of this probe which detects X and Y chromosome specific fragment lengths in human DNA offers a powerful tool in the study of human sex chromosome abnormalities and of disorders with abnormal sex differentiation. For example, individuals with either male, female, or ambiguous genitalia may be found to have one normal X chromosome and the second sex chromosome being replaced by a structurally abnormal ring or small marker chromosome of uncharacteristic banding pattern. Chromosome staining techniques are most often inadequate to determine whether such an abnormal chromosome is derived from the X or from the Y. On the other hand, individuals with a Y derived chromosome are at risk of malignant degeneration of intra-abdominal gonads or gonadal streaks. Hybridization of these patients' DNA with the *DXYS1* probe could be useful to distinguish the origin of the structurally abnormal chromosome, if it contains the region of X–Y homology.

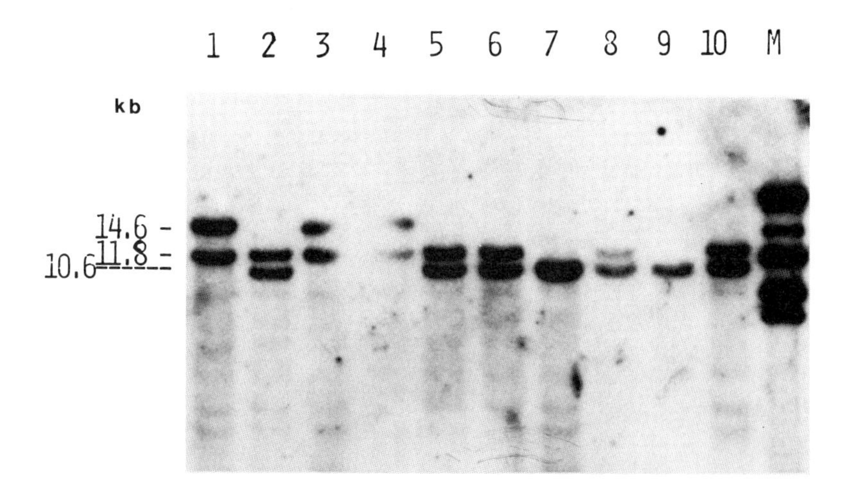

Fig. 3. Southern blot of *TaqI* digested human DNA samples (lanes 1-7 and 10) and Chinese hamster/human hybrids (lanes 8 and 9) hybridized with ^{32}P-labelled plasmid pDP31. Gel electrophoresis and transfer, and nick-translation were done as described previously.[7,9] M = molecular size markers. See text for detailed explanation of results and interpretation.

Furthermore, "sex reversed" 46,XX males are sterile. The reason for the embryonic XX gonad development into testes may be a small translocation of Y chromosomal material to the end of the short arm of the human X.[21] This Y chromosomal material would then be expected to contain the testes-determining genes. In some families X chromosomes bearing a visible translocation from the Y were passed on through carrier females. It will be of interest to look for the presence of the 14.6kb fragment in these XX males.

Preliminary data from our laboratory are shown in Figure 3. A phenotypic male who has a small isodicentric Y chromosome with two sets of short arms and a small portion of the proximal long arm does have the 14.6kb Y-specific fragment (lane 4). The pattern is the same as in normal control male DNA (lane 1). This result indicates that *DXYS1* is located either on the short arm of the Y chromosome or on the long arm close to the centromere.

The DNA from a single XX male produced both X-specific fragments and no Y-specific fragment (lane 5). The pattern is identical to that seen in heterozygous females (lanes 2, 6 and 10). The female DNA in lane 7 reveals apparent homozygozity for the 10.6kb fragment.

Human/Chinese hamster hybrids in lanes 8 and 9 are derived from a heterozygous female (lane 10). The hybrid in lane 9 had only one human X chromosome.

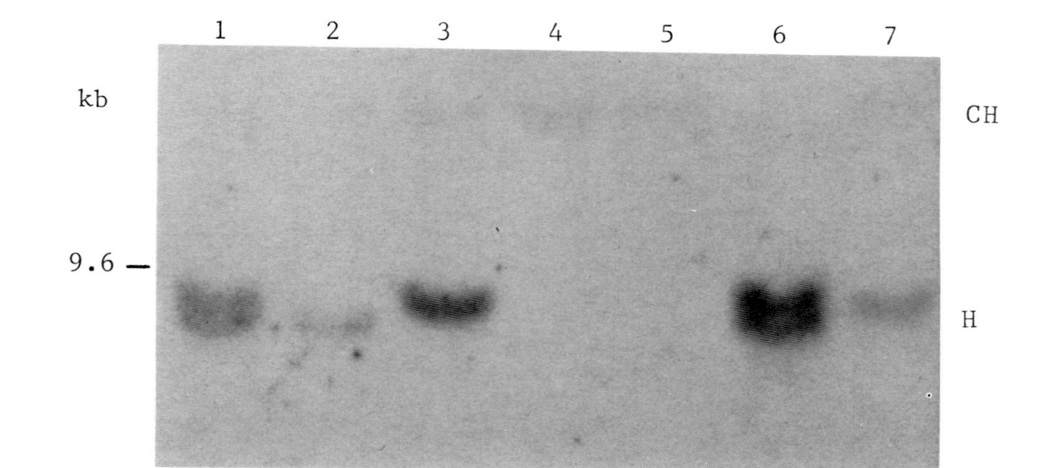

Fig. 4. Southern blot of *BglII* cleaved DNA from human donor KG7 with t(11;15) (p11;p12) translocation (lane 6), Chinese hamster parental cell line Don/a23 (lane 5) and KG7xDon/a23 hybrids (lanes 1-4 and 7) probed with ^{32}P-labelled pED.[26] Details of methods have been described elsewhere.[17] The human donor (lane 6) and the hybrid in lane 1 exhibit two restriction fragment alleles (H). A faint cross-hybridizing Chinese hamster fragment (CH) is seen in all lanes except lane 6. Hybrid in lane 1 contains the normal chromosome 11 and the derivative 15 that includes the other 11 short arm. Hybrid in lane 2 has only the normal chromosome 11. Hybrid in lane 3 has both translocation chromosomes in the absence of the normal 11. Hybrid in lane 4 has only the derivative 11 that includes the 11 long arm, and hybrid in lane 7 has only the derivative 15 with the 11 short arm.

The hybrid in lane 8 had one normal human X chromosome and the long arm of the other X involved in a rearrangement (data not shown). This rearranged chromosome was present in lower frequency and appears to contain the 11.8kb allele while the 10.6kb allele is on the structurally normal X. This very preliminary evidence suggests that the *DXYS1* sequences may be located on the long arm of the human X chromosome.

Furthermore, the samples in lanes 2 and 6 were derived from a female heterozygous for an interstitial deletion of part of band Xp21. The finding that she has both the 10.6kb and the 11.8kb alleles excludes the region that she is missing as the possible site of *DXYS1*.

HUMAN *C–HARVEY RAS* 1 ONCOGENE

In collaboration with C. Shih and R. A. Weinberg, MIT, we have studied the chromosomal location of sequences homologous to the transforming gene isolated from the EJ bladder carcinoma cell line.[22] It has since been discovered that this transforming sequence is identical to the one isolated from the T24 bladder carcinoma line by M. Goldfarb and associates,[23] because EJ and T24 are sublines derived from the same cell line. Furthermore, the bladder carcinoma transforming sequences are homologous to the viral oncogene of the Harvey murine sarcoma virus (*v-Ha-ras*).[24,25] While the viral sequence may recognize more than one sequence in human DNA, the bladder carcinoma transforming gene has been found to be homologous to the human *c-Ha-ras* 1 proto-oncogene.[24] PEJ6.6, a plasmid containing the 6.6kb *BamHI* fragment from the transforming gene, and pEC, a plasmid containing the homologous sequence isolated from normal human DNA, were both used as hybridization probes against a panel of 21 rodent/human somatic cell hybrids.[17] The hybrid DNA had been restricted with *Eco*RI or *HindIII*. The hybrids were scored for the presence or absence of the major band of hybridization. We were able to assign *c-Ha-ras* 1 to chromosome 11, with each other human chromosome being ruled out by discordancy rates of 30% or greater.[17]

In order to obtain a regional localization on chromosome 11, we turned to a series of hybrids that contained defined parts of chromosome 11. Series XXI hybrids are derived from a human male with a balanced t(11;15) (p11;p12) translocation.[27] This human donor's DNA exhibited two restriction fragments when cut with *BglII* (Figure 4, lane 6). Analysis of the hybrid DNA cleaved with *BglII* allowed us to assign the larger size fragment (Figure 4, lanes 3 and 7) to the derivative chromosome 15 containing the chromosome 11 short arm, and the smaller size fragment to the normal chromosome 11 (lane 2). The derivative chromosome 11 containing the 11 long arm did not carry sequences homologous to the transforming gene (lane 4).

In addition, hybrids from two other series that also contained 11 short arm material were positive for hybridization with pEJ6.6. All results taken together unambiguously assign the *c-Ha-ras* 1 oncogene to region 11p11→p15.[17]

This example illustrates one advantage of the somatic cell hybrid approach. It is possible to regionally map RFLP loci even in the presence of the normal homologue, as long as the hybrid DNA is cleaved with an enzyme that reveals fragment lengths polymorphism. Restriction fragments can thus be assigned to members of a homologous chromosome pair that are cytologically distinguishable, for example, by virtue of being involved in a translocation.

A number of restriction enzymes can be used to reveal polymorphism in human DNA using a *c-Ha-ras* 1 probe.[23] The assignment of this RFLP locus to the short arm of chromosome 11 places it within measurable distance of other RFLP loci: those recognized with human non-*a*-globin probes,[1] insulin,[2,3] and anonymous DNA fragments.[6]

Studying the inheritance of these RFLP in families that contain heterozygotes for more than one of these loci should provide information on the linear order and recombination fractions of these clinically relevant genes on the short arm of human chromosome 11.

The association of aniridia and predisposition to Wilms tumor is associated with heterozygous deletion of the distal half of band 11p13.[28] The locus for the autosomal dominant form of Wilms tumor has not yet been mapped. Linkage studies in Wilms tumor families might provide an RFLP marker for the detection of Wilms tumor predisposition in affected families.

CONCLUSIONS

The field is developing very rapidly. The detection and characterization of new RFLP, the physical assignment of the sequences to human chromosomes and chromosome regions, and the study of the inheritance of the RFLP alleles in families are going hand in hand. No single laboratory can be expert in all the different technologies and have equal access to DNA probes, somatic cell hybrids, *in situ* hybridization and chromosome identification expertise, and to large families with rare Mendelian disorders. Progress in the field will depend greatly on collaboration and exchange of materials and information.

ACKNOWLEDGMENTS

This work was supported by a research grant GM-26105 from the National Institutes of Health. We thank R-D Wegner, Institute for Human Genetics, FU Berlin, for human fibroblasts containing an isodicentric Y chromosome, and D. Page, MIT, for plasmid pDP31.

Figure 4 has been reproduced by courtesy of *Science*, 1983 in press; copyright 1983 by the American Association for Advancement of Science.

REFERENCES

1. Antonarakis ES, Boehm CD, Giardina PJV, Kazazian Jr. HH. Nonrandom association of polymorphic restriction sites in the *beta*-globin gene cluster. *Proc Natl Acad Sci USA* 1982; 79:137-41.
2. Rotwein P, Chyn R, Chirgwin J, Cordell B, Goodman HM, Permutt MA. Polymorphism in the 5'-flanking region of the human insulin gene and its possible relation to type 2 diabetes. *Science* 1981; 213:1117-20.
3. Bell GI, Selby MJ, Rutter WJ. The highly polymorphic region near the human insulin gene is composed of simple tandemly repeated sequences. *Nature* 1982; 295:31-5.
4. Wyman AR, White R. A highly polymorphic locus in human DNA. *Proc Natl Acad Sci USA* 1980; 77:6754-58.
5. Botstein D, White RL, Skolnick M, Davis RW. Construction of a genetic linkage map in man using restriction fragment length polymorphisms. *Am J Hum Genet* 1980; 32:314-31.
6. Skolnick MH, Francke U. Report of the committee on human gene mapping by recombinant DNA techniques. *Cytogenet Cell Genet* 1982; 32:194-204.
7. DeMartinville B, Wyman AR, White R, Francke U. Assignment of the first random restriction fragment length polymorphism (RFLP) locus (*D14S1*) to a region of human chromosome 14. *Am J Hum Genet* 1982; 34:216-26.
8. Balazs I, Purrello M, Rubinstein P, Alhadeff B, Siniscalco M. Highly polymorphic DNA site *D14S1* maps to the region of Burkitt lymphoma translocation and is closely linked to the heavy chain *gamma* 1 immunoglobulin locus. *Proc Natl Acad Sci USA* 1982; 79:7295-99.
9. Page D, *et al.* Single-copy sequence hybridizes to polymorphic and homologous loci on the human X and Y chromosomes. *Proc Natl Acad Sci USA* 1982; 79:5352-56.
10. Kunkel LM, Tantravahi U, Eisenhard M, Latt SM. Regional localization on the human X of DNA segments cloned from flow sorted chromosomes. *Nucl Acids Res* 1982; 10: 1557-78.
11. Davies KE, Young BD, Elles RG, Hill ME, Williamson R. Cloning of a representative genomic library of the human X chromosome after sorting by flow cytometry. *Nature* 1981; 293:374-76.
12. Gusella JF, *et al.* Isolation and localization of DNA segments from specific human chromosomes. *Proc Natl Acad Sci USA* 1980; 77:2829-33.
13. Lebo RV, Carrano AV, Burkhart-Schultz K, Dozy AM, Yu L-C, Kan YW. Assignment of human *beta, gamma* and *delta*-globin genes to the short arm of chromosome 11 by chromosome sorting and DNA restriction enzyme analysis. *Proc Natl Acad Sci USA* 1979; 76:5804-08.
14. Chandler ME, Kedes LH, Cohn RH, Yunis JJ. Genes coding for histone proteins in man are located on the distal end of the long arm of chromosome 7. *Science* 1979; 205: 908-10.
15. Gerhard DS, Kawasaki ES, Bancroft FC, Szabo P. Localization of a unique gene by direct hybridization *in situ. Proc Natl Acad Sci USA* 1981; 78:3755-59.
16. Nakhasi HL, Lynch KR, Dolan KP, Unterman RD, Feigelson P. Covalent modification and repressed transcription of a gene in hepatoma cells. *Proc Natl Acad Sci USA* 1981; 78: 834-37.
17. DeMartinville B, Giacalone J, Shih C, Weinberg RA, Francke U. Oncogene from human EJ bladder carcinoma is located on the short arm of chromosome 11. *Science* 1983;219: 498-501.
18. Francke U, deMartinville B. Mapping of DNA sequences to chromosome regions in somatic cell hybrids. *In*: White R, Caskey T, eds. Banbury Report: Recombinant DNA applications to human disease. (in press 1983).

19. Battey J, Max EE, McBride WO, Swan D, Leder P. A processed human immunoglobulin epsilon gene has moved to chromosome 9. *Proc Natl Acad Sci USA* 1982; 79:5956-60.
20. Francke U, Busby N, Shaw D, Hansen S, Brown MG. Intrachromosomal gene mapping in man: Assignment of nucleoside phosphorylase to region 14cen→14q21 by interspecific hybridization of cells with a t(X;14)(p22;q21) translocation. *Somat Cell Genet* 1976; 2: 27-40.
21. Evans HJ, Buckton KE, Spowart G, Carothers AD. Heteromorphic X chromosomes in 46,XX males: Evidence for the involvement of X-Y interchange. *Hum Genet* 1979; 49: 11-31.
22. Shih C, Weinberg RA. Isolation of a transforming sequence from a human bladder carcinoma cell line. *Cell* 1982; 29:161-69.
23. Goldfarb M, Shimizu K, Perucho M, Wigler M. Isolation and preliminary characterization of a human transforming gene from T24 bladder carcinoma cells. *Nature* 1982; 296:404-9.
24. Parada LF, Tabin CJ, Shih C, Weinberg RA. Human EJ bladder carcinoma oncogene is homologue of Harvey sarcoma virus *ras* gene. *Nature* 1982; 297:474-78.
25. Der CJ, Krontiris TG, Cooper GM. Transforming genes of human bladder and lung carcinoma cell lines are homologous to the *ras* genes of Harvey and Kirsten sarcoma viruses. *Proc Natl Acad Sci USA* 1982; 79:3637-40.
26. Tabin CJ, *et al.* Mechanism of activation of a human oncogene. *Nature* 1982; 300:143-49.
27. Francke U, Francke B. Requirement of human chromosome 11 long arm for replication of herpes simplex virus type I in non-premissive Chinese hamster x human diploid fibroblast hybrids. *Somat Cell Genet* 1981; 7:171-91.
28. Francke U, Holmes LB, Atkins L, Riccardi VM. Aniridia-Wilms tumor association: Evidence for specific deletion of 11p13. *Cytogenet Cell Genet* 1979; 24:185-92.

CONSTRUCTION, ANALYSIS, AND UTILIZATION OF RECOMBINANT PHAGE LIBRARIES OBTAINED USING FLUORESCENCE ACTIVATED FLOW SORTING

Samuel A. Latt
Louis M. Kunkel
Umadevi Tantravahi
Jeff Aldridge
Marc Lalande

INTRODUCTION

Fluorescence activated chromosome sorting (Figure 1) facilitates the acquisition of chromosome-specific DNA fragments which serve as probes in the study of human chromosome structure and function. Techniques for metaphase chromosome isolation[2-4] and staining[5,6] now permit differentiation according to chromosome size and/or composition of most of the human karyotype. A few hundred thousand chromosomes per hour, enriched 10–20 fold for specific chromosomes or subchromosomal segments, can be isolated by flow sorting. DNA from these chromosomes can be prepared in a form suitable for cloning in *lambda* phage vectors (with typical yields in the range of 100,000 p.f.u. per 100 ng DNA), and the cloned inserts screened effectively for localization to the regions of interest.[7-9] Large numbers of DNA fragments from specific regions of a chromosome become particularly useful in applications, *e.g.* linkage analysis or exploration of contiguous sequences, requiring multiple segments in proximity to each other.

Structurally abnormal chromosomes facilitate both chromosome isolation and sequence screening. Conversely, possession of a large number of sequences localized to given chromosome segments provides a grid for reinforcing cytotological analyses of abnormal chromosomes, for linkage analysis, and for molecular studies of functional differences along chromosomes, such as the human X.

RESULTS AND DISCUSSION

Our laboratory has utilized fluorescence activated chromosome sorting to study two types of problems. The first was concerned with mouse and human X chromosome structure and function[8,10] and the second involved amplified DNA

ISBN 0-12-492220-1

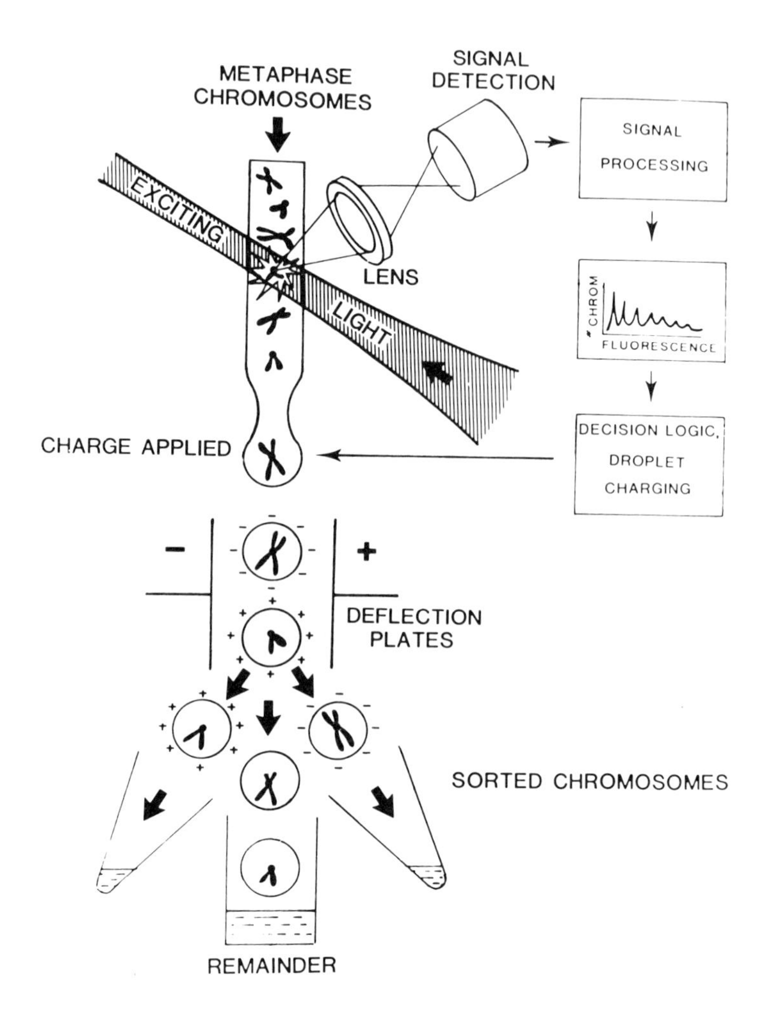

Fig. 1. Diagrammatic representation of fluorescence activated metaphase chromosome analysis and sorting (adapted from Melamed *et al*).[1]

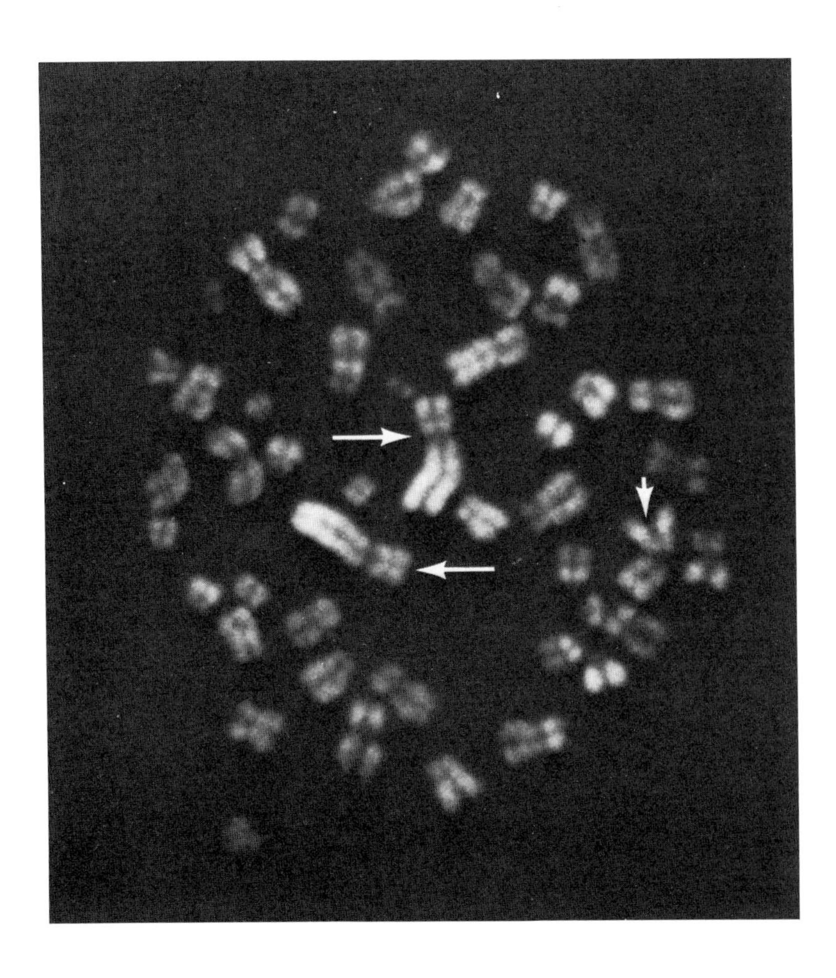

Fig. 2. Metaphase chromosomes from an IMR–32 cell substituted with BrdU and stained with 33258 Hoechst to highlight early replicating chromosomal regions.[13] Long arrows point to the HSR containing #1 chromosomes and a short arrow to the normal #1.[11]

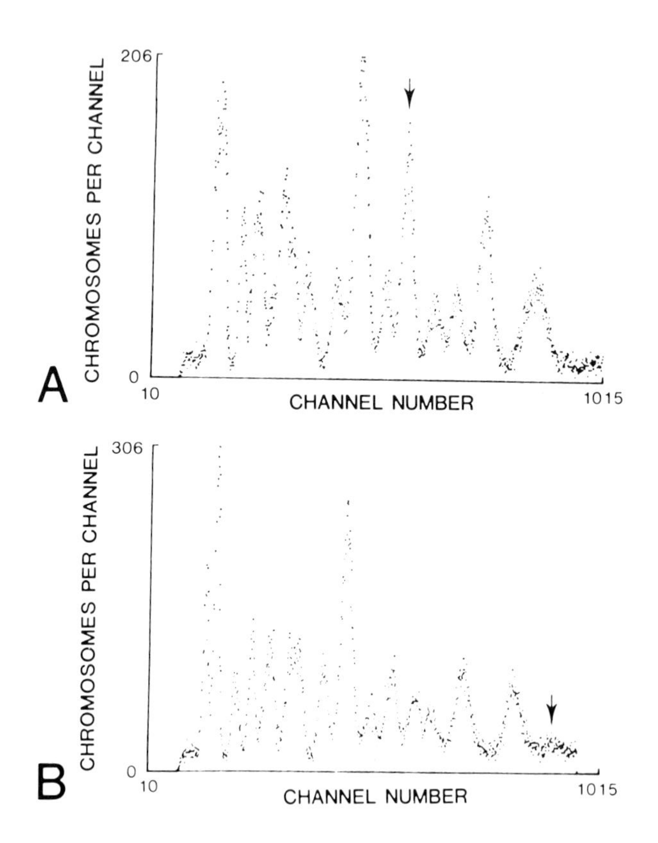

Fig. 3. DNA flow histograms of metaphase chromosomes from human lymphoblastoid cells containing supernumerary and structurally abnormal X chromosomes. (A) 49, XXXXY cells and (B) 46,X, dic (X) (q24) cells were utilized. Chromosomes were isolated by the method of Sillar and Young[4] and stained with 33258 Hoechst (Lalande *et al*, in preparation). The peak containing the X (in A) and the dic (X) (in B) are indicated by arrows.

---

sequences, such as those present in homogeneously staining regions (HSR) of human IMR-32 neuroblastoma cells. The latter work, initially reported at a Banbury Conference in 1982[11] and described in detail elsewhere[12] (and Kanda *et al*, in preparation), permitted the isolation of sequences within the HSR of IMR-32 (Figure 2). These latter sequences are being used to characterize the mobility, degree of repetition, expression, and perhaps the biological significance of amplified DNA in this and other neuroblastoma-derived tissue.

In former experiments, a size difference was exploited to enrich for the X[7] chromosome from cells carrying the mouse Cattanach translocation T(X;7)Ct,[10] and a numerical difference was utilized to facilitate enrichment for the human X[8]

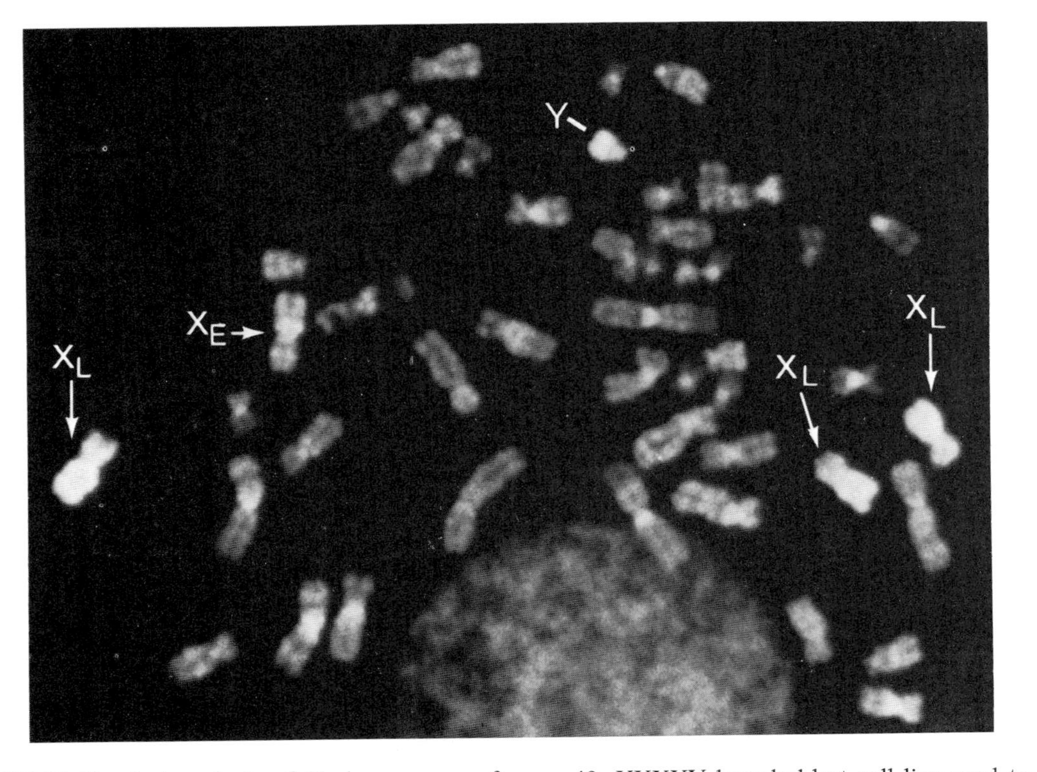

Fig. 4. BrdU–33258 Hoechst analysis of X chromosomes from a 49, XXXXY lymphoblast cell line used to obtain the chromosomes shown in Figure 3A. Bright fluorescence highlights late replication; long arrows point to the late replicating X and a short arrow points to the early replicating X.

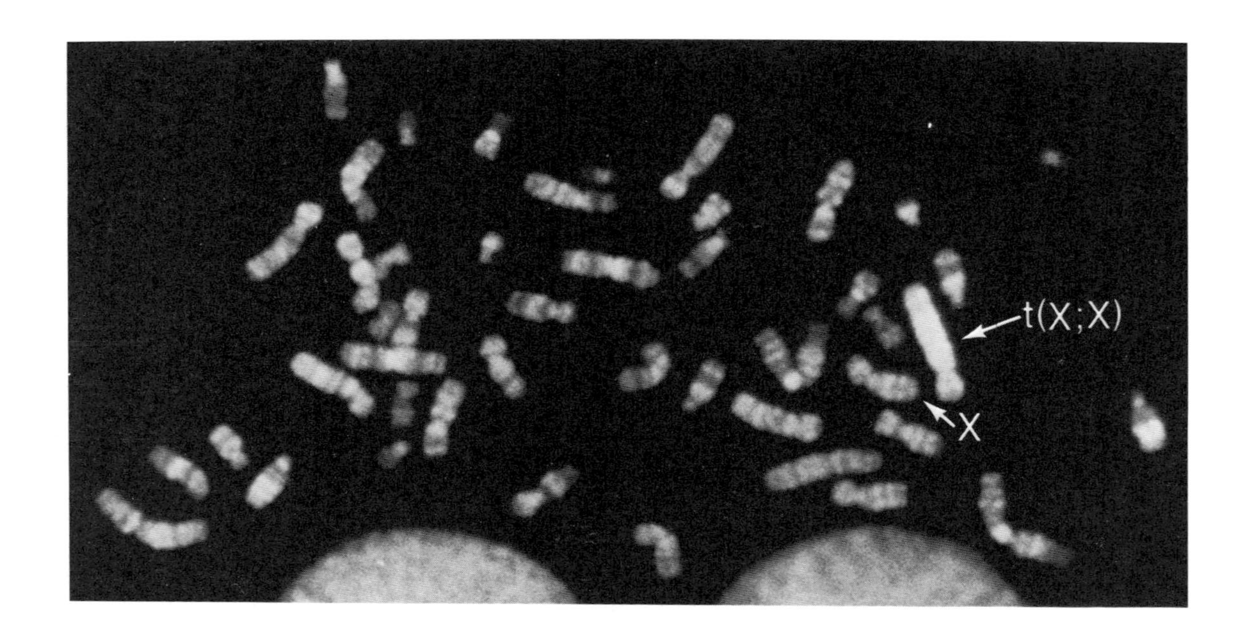

Fig. 5. BrdU–33258 Hoechst analysis of X chromosomes from the 46 X, dic(X) (q24) cell line used to obtain the chromosomes shown in Figure 3B. The long arrow points to the dicentric X, while the short arrow points to the normal, early replicating X.

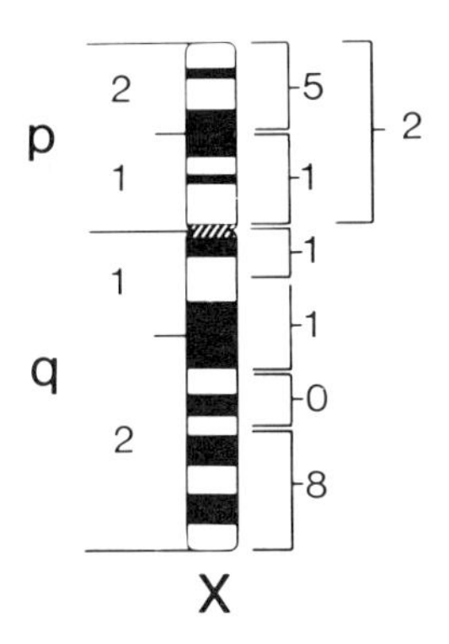

Fig. 6. Diagrammatic localization of some cloned DNA fragments assigned in our laboratory to the human X chromosome. Many of the fragments in Xp are being localized in collaborative experiments with U. Francke and B. deMartinville.

in a manner analogous to the experiments of Davies *et al.*[7] More recently, we have been employing a modification of the Sillar and Young procedure[4] (utilizing 33258 Hoechst rather than ethidium bromide as a stain) for high efficiency analysis and sorting of both normal (Figure 3A) and structurally abnormal (Figure 3B) human X chromosomes (Figures 4,5). From the former, phage libraries with larger inserts and representing the X more completely than was possible with previously prepared phage libraries are being prepared which contain small, total *HindIII* or *Eco*RI endonuclease digest inserts. From the latter, libraries enriched for subchromosomal regions are being prepared.

Thus far, by screening repeat negative DNA segments obtained from flow sorted chromosome libraries, as well as from a lesser number of cDNA homologous clones, we have identified more than 20 human X-specific inserts. Of these 18 have been mapped along the X using Southern blot hybridizations.[14] These hybridization analyses utilize DNA isolated either from human-rodent hybrids retaining all or part of the human X, or from human cell lines containing supernumerary or structurally abnormal X chromosomes (Figure 6)[8] (collaborative studies with U. Francke *et al*, in progress). It is not clear whether the apparent clustering of probes at the termini of Xp and Xq is influenced by the method used for probe selection or simply reflects fluctuation associated with small sample size. The former possibility would imply the existence of a non-random relative distribution of *Alu*-like repeats and *HindIII* sites along the human X chromosome. Studies examining this point are in progress.

One application of this collection of probes, which depends in part on quantitative analysis of Southern blot data to differentiate between representations of one and two copies of a given sequence per genome, has been the confirmation (Tantravahi *et al*, in preparation) that a human X chromosome structural abnormality from a patient with Turner's syndrome reflected a deletion

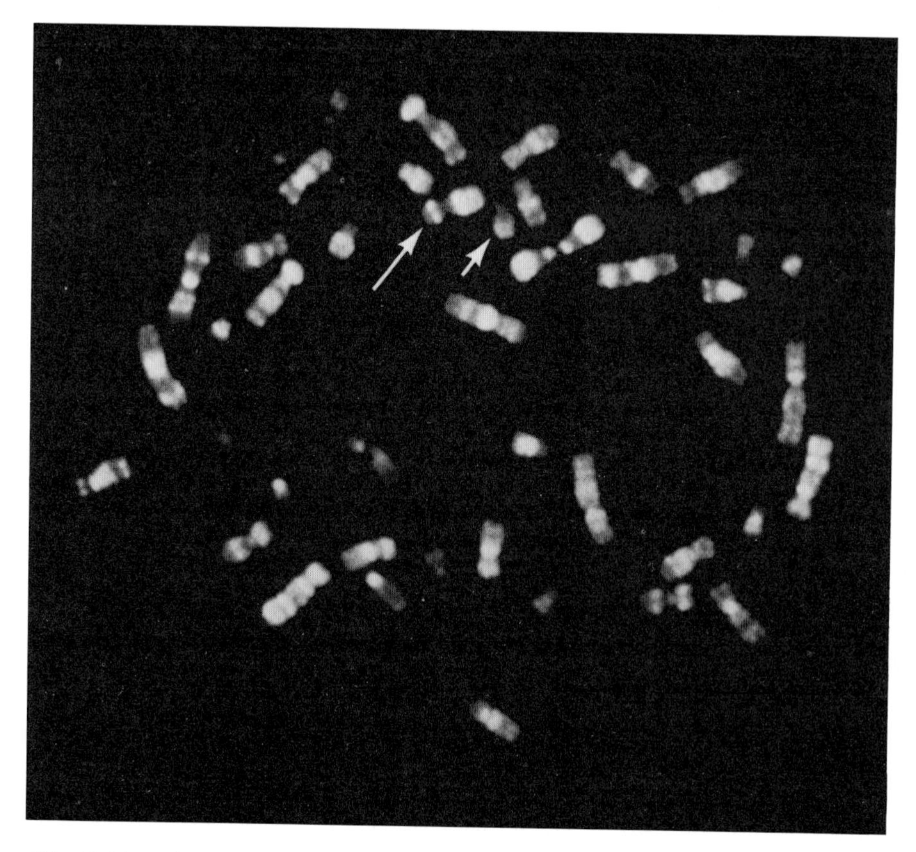

Fig. 7. Metaphase chromosomes from a 46, XXq– cell, stained as described in the legends to Figures 4 and 5 to highlight late replicating chromosomal regions. The normal X is indicated by a long arrow and the deleted X by a short arrow.

---

of most of the long arm of the X (Figures 7, 8A). One other probe, previously localized to the long arm of the X,[8] was found to be present in at least a two copy level in cells from this patient (Figure 8B) (Tantravahi *et al*, in preparation), placing it in or near the region at the proximal part of X_q that has been hypothesized by some (*e.g.* ref. 15) to be crucial for X chromosome inactivation. The deletion mapping illustrated by Figure 8 should be of use to investigators attempting to localize sequences within regions of other chromosomes, *e.g.* 11p13 or 13q14, which are often deleted in human neoplasms, Wilms tumor,[16] or retinoblastoma,[17] respectively. The cloned sequence mapping near the putative X chromosome inactivation center (Figure 8B), which by RNA blotting, does not appear to be appreciably transcribed (U. Tantravahi, unpublished data), is now being employed in an analysis of DNA and chromatin organization in this region.

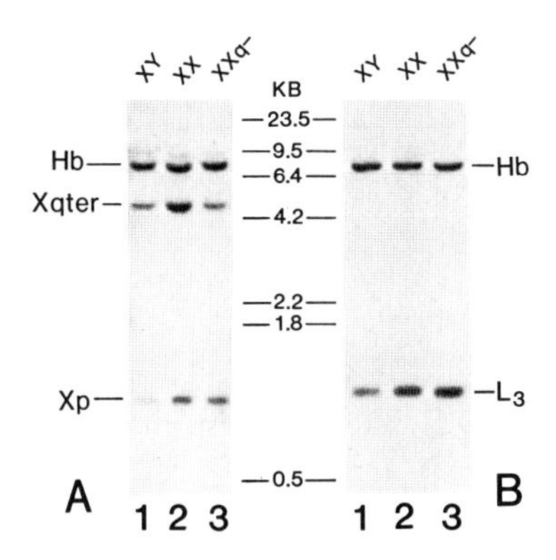

Fig. 8. Southern blotting of cloned X-specific probes to DNA from a cell with a deletion of most of the long arm of the human X. The center indicates sizes determined from a ^{32}P-labelled *HindIII* digest of phage *lambda* DNA as a size standard. Lanes 1, 2, and 3 contain equal amounts of DNA from 46XY, 46XX, and 46X,Xq– human cells respectively. (A) Comparative hybridization of probes from Xp and Xq24→Xqter. The former hybridizes twice as intensely to 46XX and 46,X,Xq– DNA than to 46,XY DNA, while the latter shows equal hybridization to the 46,XY and 46,XXq– DNA, as expected if it is present in the segment deleted from Xq. Hybridization of a beta globin probe, equally to all three DNA samples, is shown for comparison. (B) Localization of an X-specific probe to a small region between X(cen) and Xql. This probe, which localizes to Xq, based on hybridization intensity to DNA from 46,X,i(Xq) cells, hybridizes to 46,X,Xq– DNA at least as intensely as to 46, XX, DNA, placing it in the very proximal part of the long arm of the human X. Visual assessments are substantiated by densitometric analysis (Tantravahi *et al*, in preparation).

An additional cloned DNA segment with potentially interesting biological properties, isolated from our flow-sorted X phage library, homologous to a HeLa cDNA clone and localized to the short arm of the X, hybridizes with DNA sequences present on the human Y and is being utilized to study homology between the human X and Y chromosomes (Kunkel *et al*, in preparation). Another DNA segment with both X and Y homology has recently been described by Page *et al.*[18]

A final illustration of the use of cloned probes from flow sorted libraries is the identification of putative restriction fragment length polymorphisms (RFLP)[19,20] within a subregion of the human X. The search for RFLP's among these fragments, for use in linkage analysis, has already identified two candidates which are located in the distal part of the long arm of the X (Xq24→Xqter) (Aldridge, Kunkel *et al*, unpublished data) (Figure 9). Depending on assumptions comparing linkage map distances with physical distances and on the effective heterozygosity of regions sampled by different probes, perhaps 25-50 such probes located throughout the X will be needed. These will comprise a grid of specific DNA fragments relative to which individual genetic loci, *e.g.* that for Duchenne muscular dystrophy for which a specific DNA probe does not yet exist, can be mapped. Preparation of the flow sorted libraries from limited regions of the X, such as the Xpter→Xql fragment described in Figure 7, may prove to be particularly valuable for this latter specific goal, in view of the presumed localization of a genetic locus responsible for Duchenne muscular dystrophy to the short arm of the human X.[21,22]

Experiments made possible by large numbers of human X-specific probes include those directed at correlating the dosage compensation of gene expression, late replication, and chromatin organization as examined by nuclease digestion in different types of cells. More immediate, methodological questions concern subsequent strategies to be employed in extending results obtained or easily obtainable with present phage libraries of flow sorted chromosomes, consisting of small DNA inserts from total *Eco*RI or *HindIII* digests. Complete libraries of larger phage inserts preparered, for example, from partial digestion of chromosomal DNA with *MboI* endonuclease and cloned in the phage *Charon* 30[23] or in cosmids[24] are needed. However, it is not clear whether these libraries are more efficiently obtained from total genomic DNA of normal or hybrid[25] cells, from DNA partially enriched for groups of chromosomes by low G sedimentation,[26] which would then be screened with existing probes from flow sorted chromosome-enriched libraries, or from DNA of chromosomes isolated by more exotic methodology[27] employing microdissection. A second problem is the need for methods for cross-screening different types of X-enriched libraries, to speed up the identification of X-specific sequences and to search for sequences conserved between different species. A third class of experiments involves utilization of flow sorting to isolate chromosomes for DNA fragment localization, as initially done by Lebo *et al,*[28] an approach made more accessible by continued technological improvements. These and related questions are becoming increasingly plausible options which will probably be exercised to different degrees in coordinate molecular and cytological studies of human chromosomes to obtain information of both basic and clinical significance.

SUMMARY

Fluorescence activated sorting of metaphase chromosomes has proved useful for obtaining recombinant DNA libraries enriched for specific chromosomes or subchromosomal regions. Different aspects of this approach are outlined and

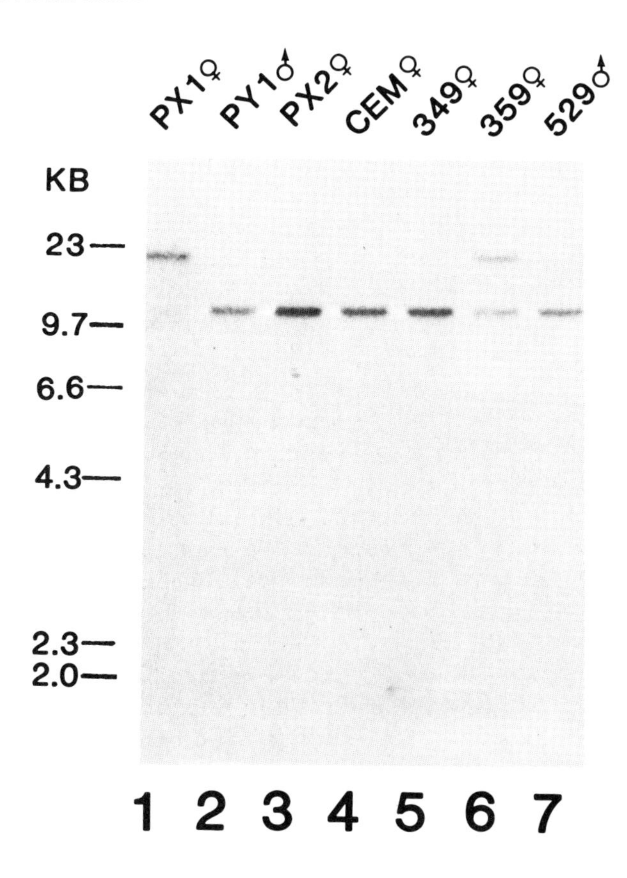

Fig. 9. Hybridization of a cloned DNA segment from X q24–Xqter in a manner expected for a restriction fragment length polymorphism. DNA from seven individuals was digested with *TaqI* endonuclease. DNA samples were obtained from CCRF-CEM cells (lane 4) or placentas (other lanes). Note fragments approximately 20 KB in size in lanes 1 and 6; fragments of approximately 10 KB are seen in all lanes (albeit faintly in lane 1). Indicated at the left are sizes of Hind III digest fragments of phage *lambda* DNA. It should be noted that the diversity of hybridization patterns detected by the probe in this figure only putatively reflects polymorphism. To establish this conjecture more firmly, the inheritance of the variant hybridization patterns within individual pedigrees will have to be demonstrated. Experiments addressing this point are in progress, and the results thus far are consistent with the provisional assumption made.

various applications of one such library, enriched for the human X chromosome, is described. Possession of large numbers of cloned DNA segments localized to different regions of the human X chromosome is shown to be useful as an adjunct to cytological characterization of structurally abnormal X chromosomes and as a potential source of probes needed for constructing a linkage map of the human X. Comparison of the expression and chemical state of such segments on the X should ultimately also be helpful in a molecular analysis of X chromosome structure, replication, and function.

ACKNOWLEDGMENTS

The technical assistance of Michael Eisenhard, Louis Juergens, and Alan Flint is greatly appreciated. This research was supported by grants from the National Institutes of Health (GM21121, HD04807) and the National Foundation March of Dimes (1–353). Louis M. Kunkel is the recipient of a fellowship from the Muscular Dystrophy Association of America. A similar version of the present manuscript has been written for publication in the proceedings of an October 1982 Banbury Conference on Recombinant DNA Applications to Human Disease. Some figures reprinted from "Gene Amplification", Cold Spring Harbor Laboratory and "Genetic Analysis of the X Chromosome", Plenum Press, 1982, p 50.

REFERENCES

1. Melamed MR, Mullaney PF, Mendelsohn ML. Flow Cytometry and Sorting. New York: John Wiley, 1979.
2. Carrano AV, Gray JW, Langlois RJ, Burkhart-Schultz KJ, Van Dilla MA. Measurement and purification of human chromosomes by flow cytometry and sorting. *Proc Nat Acad Sci USA* 1979; 76:1382-4.
3. Otto FJ, Olidges H, Gohde W, Barlogie B, Schumann J. Flow cytogenetics of uncloned and cloned Chinese hamster cells. *Cytogenet Cell Genet* 1980; 27:52-6.
4. Sillar R, Young BD. A new method for the preparation of metaphase chromosomes for flow analysis. *Histochem Cytochem* 1981; 29:74-8.
5. Latt SA. Fluorescent probes of DNA microstructure and synthesis. *in*: Melamed MR, Mullaney PF, Mendelsohn ML. eds. Flow Cytometry. New York:John Wiley, 1979: 263-84.
6. Gray JW, Langlois RG, Carrano AV, Burkhart-Schultz K, Van Dilla MA. High resolution chromosome analysis: one and two parameter flow cytometry. *Chromosoma* 1979; 73:9-27.
7. Davies KE, Young BD, Elles RG, Hill ME, Williamson R. Cloning of a representative genomic library of the human X chromosome after sorting by flow cytometry. *Nature* 1981; 283:374-6.
8. Kunkel LM, Tantravahi U, Eisenhard M, Latt SA. Regional localization on the human X of DNA segments cloned from flow sorted chromosomes. *Nucl Acids Res* 1982; 10: 1557-68.
9. Krumlauf R, Jeanpierre M, Young BD. Construction and characterization of genomic libraries from specific human chromosomes. *Proc Nat Acad Sci USA* 1982; 79:2971-75.
10. Disteche CM, Kunkel LM, Lojewski A, Orkin SH, Eisenhard M, Sahar E, Travis B, Latt SA. Isolation of mouse X-chromosome specific DNA from an X-enriched lambda phage library derived from flow sorted chromosomes. *Cytometry* 1982; 2:282-6.

11. Latt SA, Alt FA, Schreck RR, Kanda N, Baltimore D. The use of chromosome flow sorting and cloning to study amplified DNA sequences. *in*: Schimke RT, ed. Gene Amplification. Cold Spring Harbor Laboratory:Long Island, 1982: 283-9.
12. Kanda N, Schreck RR, Alt FW, Baltimore D, Latt SA. Use of fluorescence activated metaphase chromosome flow sorting to facilitate the isolation of cloned DNA from the HSR of IMR-32 cells. *Amer J Hum Genet* 1982; 34:131A.
13. Latt SA. Microfluorometric detection of DNA synthesis in human metaphase chromosomes. *Proc Nat Acad Sci USA* 1973; 70:3395-99.
14. Southern EM. Detection of specific sequences among DNA fragments separated by gel electrophoresis. *J Mol Biol* 1975; 48:503-17.
15. Therman E, Sarto G, Palmer D, Kallio H, Denniston C. Position of the human X inactivation center on Xq. *Hum Gen* 1979; 50:59-64.
16. Francke U, Holmes LB, Atkins L, Riccardi VM. Aniridia-Wilm's tumor association; evidence for specific deletion of 11q13. *Cytogenet Cell Genet* 1979; 24:185-92.
17. Knudson AG, Meadows AT, Nichols WW, Hill R. Chromosomal deletion and retinoblastoma. *N Engl J Med* 1976; 295:1120-23.
18. Page D, De Martinville B, Barker D, Wyman A, White R, Francke U, Botstein D. Single-copy sequence hybridizes to polymorphic and homologous loci on human X and Y chromosomes. *Proc Nat Acad Sci USA* 1982; 79:5352-56.
19. Botstein D, White RL, Skolnick M, Davis RW. Construction of a genetic linkage map in man using restriction fragment length polymorphisms. *Amer J Hum Genet* 1980; 32: 314-31.
20. De Martinville B, Wyman AR, White R, Francke U. Assignment of the first random restriction length polymorphism (RFLP) locus (D1451) to a region of human chromosome 14. *Amer J Hum Genet* 1982; 34:216-26.
21. Jacobs PA, Hunt PA, Bart RD. Duchenne muscular dystrophy (DMD) in a female with and X/autosome translocation: further evidence that the DMD locus is at X_p21. *Amer J Human Genet* 1981; 33:531-9.
22. Murray JM, Davies KE, Harper PS, Meredith L, Mueller CR, Williamson R. Linkage relationship of a cloned DNA sequence on the short arm of the X chromosome to Duchenne muscular dystrophy. *Nature* 1982; 300:69-72.
23. Williams BG, Blattner FR. Bacteriophage *lambda* vectors for DNA cloning. *In:* Setlow J. Hollaender A, eds. Genetic Engineering. Vol. 2. New York:Plenum, 1980:201-81.
24. Grosveld FG, Dahl H-HM, deBoer E, Flavell RA. Isolation of beta-globin-related genes from a human cosmid library. *Gene* 1981; 13:227-37.
25. Gusella J, Keys C, Varsanyi-Breiner A, Kao FT, Jones C, Puck TT, Housman D. Isolation and localization of DNA segments from specific human chromosomes. *Proc Nat Acad Sci USA* 1980; 77:2829-33.
26. Collard JG, Schijven J, Tulp A, Meulenbroek M. Localization of genes on fractionated rat chromosomes by molecular hybridization. *Exper Cell Res* 1982; 137:463-9.
27. Scalenghe F, Turco E, Edstrom JE, Pirotta V, Melli M. Microdissection and cloning of DNA from a specific region of Drosophila melanogaster polytene chromosomes. *Chromosoma* 1981; 82:205-16.
28. Lebo RV, Carrano AV, Burkhart-Schultz K, Dozey AM, Yu LC, Kan YW. Assignment of human beta, gamma, and delta globin genes to the short arm of chromosome 11 by chromosome sorting and DNA restriction enzyme analysis. *Proc Nat Acad Sci USA* 1979; 76:5804-08.

USE OF X CHROMOSOME PROBES TO SEARCH FOR THE MOLECULAR BASIS OF X CHROMOSOME INACTIVATION

Barbara R. Migeon
Stanley F. Wolf
Ethylin W. Jabs

The mammalian X is unique among chromosomes. Not only does it harbor at least 150 loci coding for the same gamut of gene products that are specified by autosomes, but it also contains loci necessary to maintain oocytes in the female gonad. Although little is known about the nature of these important sex related gene(s), it is clear that a double dose is required in the female germ cell, but more than one X chromosome is detrimental in the male. As a consequence the regulation of the X chromosome in somatic cells of the mamalian female is complex. Superimposed upon all the mechanisms that regulate activity of autosomal genes are mechanisms to compensate for the X dosage difference between the sexes, resulting in a single active X chromosome in both sexes.[1]

It is likely that equalization of X gene products between males and females is attributable to multiple events. The initial event, occurring about the time of implantation in most species, is unknown, but results in inactivation of all but a single X in diploid cells. Yet inactivation may not be the primary event, as all but a single X are inactivated in diploid cells with multiple X chromosomes and *no* chromosome is inactivated in triploid cells of 69 XXY karotype.[2] The ratio of X chromosomes to autosomes seems to determine the number of *active* X chromosomes so that the initial step may be maintenance of X activity in response to a factor of autosomal origin.

Once inactivated, the X chromosome remains inactive in somatic cells. This implies either that the mechanisms involved in the initial event can maintain the inactivity (or activity) from one cell to its progeny, or that other mechanisms have evolved to do so.

On the other hand, any mechanism for the maintenance of X inactivity must explain how the process can be reversed so that the entire X chromosome can be reactivated during ontogeny of the female germ cell.[3,4] Further complexity of X chromosome regulation is apparent when one considers evolutionary aspects of the

ISBN 0-12-492220-1

process. What is the origin of the mammalian X and how were genes on this chromosome – many of them comparable in function to genes on any autosome – regulated prior to inactivation? Having a single active X in mammalian cells of both sexes serves to equalize expression of X linked genes in male and female somatic cells, but results in an effective monosomy for these genes. It is not clear why the monosomy for X genes in mammalian cells is so well tolerated while monosomy of autosomal genes is not.

Further evidence for more than one level of regulation underlying X chromosome activity is our observation of differential expression of loci on active and inactive X. It is clear that alleles on the inactive X in normal human cells that escaped inactivation in development, or have been derepressed, produce less gene product than those on the active X. We have analyzed unique cells in which the locus for glucose 6 phosphate dehydrogenase (*G6PD*) on the inactive X is expressed,[5] as well as cells characteristic of all human females that express the steroid sulfatase (*STS*) locus on the inactive X.[6]

In each case it is clear that the locus on the inactive X has less activity than the corresponding locus on the active homologue. The reduced expression of the *G6PD* A and wild type *STS* alleles on the inactive X may result from reduced transcription associated with the mutational event that resulted in derepression, or the influence of neighboring inactive chromatin. On the other hand, the reduced expression of these loci on the inactive X could reflect inherent differences between the two X chromosomes in levels of transcriptional activity for loci that are expressed on both chromosomes. The greater activity of loci on the active X that we observed could represent enhanced transcription of genes on that chromosome to compensate for the single copy of X linked genes.

The Search for Differences Between Active and Inactive X Chromosomes

Although there are striking differences between X homologues with respect to timing of DNA replication[7,8] and condensation in interphase, the difference at metaphase has been more subtle. Banding patterns are essentially similar and the only significant cytological differences observed has been relative resistance to hot KCL treatment[9] (Migeon manuscript in preparation). Until recently there has been little opportunity to study the molecular organization of the two chromosomal proteins, sequence arrangements or DNA modifications.

Although one expects to find differences between the two X homologues, at the molecular level, it is even more likely that both chromosomes differ from autosomes in at least some aspects of molecular organization. Little is known about the special features of the X chromosome that make it subject to this unique kind of regulation. The availability of cloned fragments of human X chromosome DNA provide a means to explore the organization of the human X. Sequences on the X which are shared with autosomes may be informative in regard to features common to all chromosomes; on the other hand, sequences unique to the X could be used as hybridization probes to reveal characteristics of X chromosome organization underlying the inactivation process. Furthermore, X specific sequences should facilitate the isolation of the inactive X chromosome as a Barr body.

Isolation of X Specific Fragments

Toward this end we developed a method of identifying and isolating X chromosome DNA sequences from a cloned human DNA library. We reasoned that Southern blotting could be used to identify recombinant plasmids containing X chromosomal DNA sequences. Plasmids could be characterized by the pattern they produced when hybridized to DNA from mouse-human hybrid cells containing the X chromosome as the sole human chromosome, and to DNA from derivative cells lacking the relevant X chromosome. If there were significant homology between the human sequences in the probe and those in the mouse DNA, we expected that hybridization with DNA from hybrid cells would occur regardless of the presence of the human X. However, human sequences in the probe that hybridize only to the hybrids with the human X are likely to be on the human X chromosome.

Recombinant plasmids with X chromosome sequences (X DNA) could be further evaluated to identify those that are found only on the human X. Therefore, the second step is to hybridize plasmids containing X DNA sequences to Southern blots of digested DNA from cells differing in X chromosome content. Plasmids containing sequences shared between the X and autosomes should hybridize equally well with either XY or XXX DNA, while those containing X specific sequences would show a dosage effect.

Using this strategy, we have been able to identify clones that contain single copy sequences unique to the human X as well as clones having sequences shared with autosomes.[10] By screening a partial library derived from *BamHI* digested 47 XXX human DNA inserted into pBR322, two recombinant plasmids (307 and 604) have been identified that have sequences that come from the X and are found only on the X. These have been mapped to Xq22-Xq25 using DNA from somatic cell hybrids and Southern blotting.

Studies of X DNA Methylation

Recent studies by others have suggested that DNA methylation is involved in regulating X chromosome expression.[11-13] The availability of X chromosome specific probes enable us to study the methylation of X chromosome DNA. The methyl sensitive restriction enzyme, *HpaII*, will not cut DNA if the sequence it recognizes is methylated; therefore, it can be used as an assay for X DNA methylation in the regions for which DNA probes are available. We have compared active and inactive X DNA methylation around X probes 307 and 604 in DNA from placentas, clonal fibroblast cultures, and mouse-human hybrid cells retaining human X chromosomes.[14,15] These studies show that in the 22 kb region around our random probes:

1. There are no obvious differences in DNA methylation between active and inactive X.
2. Methylation is not stable in primary fibroblasts.
3. Human X DNA methylation is more stable in the mouse-human hybrid cell.
4. The active X can be highly methylated while the inactive X can be less methylated.

Although our observations do not preclude important differences in DNA methylation between the X homologues, they do indicate that these differences are not ubiquitous. In any event, the changing patterns of methylation that we observed raise some questions about the fidelity with which methylation could maintain a differentiated state. Even if methylation is involved in maintaining X chromosome inactivity, it most likely is only one of several components that influence transcriptional activity. For example, the accessibility of DNA to methylases and the subsequent methylation at particular sites may depend on chromatin structure. The more stable methylation pattern for the human X in interspecies hybrid cells may reflect changes in the chromatin structure attributable to a foreign environment, or species differences in relevant determinants.

The Search for DNA Sequence Rearrangements

Using X specific random probes 307 and 604 and the 3′ cDNA sequence of *G6PD*,[16] we have searched for differences in lengths of restriction fragments between active and inactive X; such differences would be expected if sequences on the inactive X had been rearranged. The methods that we used are sensitive enough to detect sequence rearrangements that occur at frequent intervals and involve hundreds of base pairs, but not able to detect small differences at a limited number of critical sites. Southern blot analysis of DNA from 690 X chromosomes digested by at least one of 18 restriction enzymes did not reveal any consistent difference between male and female specimens or active and inactive X. Therefore, we obtained no evidence for frequent rearrangements along the length of the inactive chromosome.

Isolation of the Barr Body

The sex chromatin mass in interphase cells provides an excellent source of inactive X chromosomes for study. Unfortunately, it has been difficult to purify the Barr body, primarily because of inability to identify it during the purification process.

The availability of X specific DNA provides the necessary probes to facilitate purification of inactive X DNA. These probes can also be used to identify inactive X chromatin, if the isolation buffers are such that chromosomal proteins are preserved. The inactive X can be identified in DNA prepared from various fractions obtained during the purification. Specific identification of enriched X DNA is based on comparisons of hybridization to X probes and autosomal ones, or by utilization of restriction fragment length polymorphisms that can distinguish active from inactive X in heterozygous cells.

We are using probe 307 and a *PstI* polymorphism to determine the subcellular localization and special features of inactive X chromatin as preliminary steps to purification of the inactive X chromosome.

The Search for Reiterated and Conserved DNA Sequences Specific to the Mammalian X

The molecular basis for chromosome imprinting remains to be determined. One possible mechanism might be sequences repeated many times on the X but on no other chromosome. Such sequences might also be highly conserved with those on X chromosomes of other mammalian species.

Because the X enriched partial library that we have screened contains a good collection of reiterated sequences represented on the X, we are trying to identify those sequences that may play a role in X inactivation.

We can identify relevant sequences by characterizing each cloned fragment with regard to relative homology to DNA on the human X and autosomes, and to sequences on mouse chromosomes as well. This can be done most efficiently by a screening procedure utilizing dot blot methodology.[17] DNA from plasmids can be dotted on replicate filters and these filters can be hybridized with one of the following labeled probes: total genomic DNA, X+ hybrid DNA, X– hybrid DNA and mouse parent DNA, and the resultant autoradiographs can be compared.

We have screened 101 recombinant plasmids in our partial library using ^{32}P total genomic human DNA and classified these with respect to sequence frequency based on intensity of hybridization compared to sequences of known copy number. Among the 72 moderate or highly reiterated sequences that we identified in the plasmid library, at least 20 were found to be on the X chromosome. However, none of these were X specific, but two were significantly enriched on the X chromosome. The lack of X chromosome specific reiterated sequences may reflect the relatively small number of clones analyzed, but their absence among a significant number of such sequences indicates that most repeated sequences that are on the X are also found on autosomes. Although 75% of the human reiterated sequences hybridized with mouse DNA, none was as highly reiterated in the human genome. This suggests that many, if not most, highly reiterated sequences are species specific. (Jabs EW, Wolf SF, Migeon BR., unpublished)

Although we have sampled only a small fraction of X chromosomal DNA, there are several interesting observations regarding the composition of the human X chromosomes. Despite the conservation of gene content in X chromosomes of all mammals, we find that none of the human DNA sequences that hybridize to the human X chromosome hybridize with the mouse genome to the same extent. On the other hand, there is a significant amount of sequence homology between human autosomes and the human X chromosome. It seems that the X chromosome of one mammal may be considerably different from the X of another, and less different from autosomes than might be expected. Therefore, the features that make this chromosome susceptible to inactivation in somatic cells are relatively subtle ones.

The random probes that we have used to explore the molecular organization of the human X chromosome have been extremely useful in revealing not only the significant similarities between X and autosomes, but also the lack of obvious differences between the X homologues. Our studies with these probes indicate that there are no ubiquitous differences in methylation between active and inactive X, nor obvious X specific highly reiterated sequences. Therefore, if differences do exist, they must be limited to critical regions of the X chromosome. It may be that the search for these special regions of the chromosome will require specific gene probes that can reveal the relevant DNA sequences at these critical sites. The availability of cDNA probes for *G6PD* and *HPRT* should facilitate these studies.

On the other hand, random probes will continue to be useful as a means to identify the presence of X chromosomes in cellular DNA, for mapping, for isolating fractions of the cell enriched for X DNA, and for studies of chromatin structure.

ACKNOWLEDGMENTS

These studies were supported by National Institutes of Health grant HD05465. E.W. Jabs is a fellow of the Stetler Foundation.

REFERENCES

1. Lyon MF. X chromosome inactivation and developmental patterns in mammals. *Biol Rev* 1972; 47:1-35.
2. Migeon BR, Sprenkle JA, Do TT. Stability of the "two active X" phenotype in triploid somatic cells. *Cell* 1979; 18:637-41.
3. Gartler SM, Rivest M, Cole RE. Cytological evidence for an inactive X chromosome in murine oogonia. *Cytogenet Cell Genet* 1980; 28: 203-207.
4. Kratzer PG, Chapman VM. X chromosome reactivation in oocytes of Mus caroli. *Proc Nat Acad Sci USA* 1981; 78:3093-97.
5. Migeon BR, Wolf SF, Mareni C, Axelman J. Depression with decreased expression of the G6PD locus on the inactive X chromosome in normal cells. *Cell* 1982; 29:595-600.
6. Migeon BR, Shapiro JJ, Norum RA, Mohandas T, Axelman J, Dabora RL. Differential expression of the steroid sulfatase locus on the active and inactive human X chromosome. *Nature* 1982; 299:838-40.
7. German JL. III DNA synthesis in human chromosomes. *Trans NY Acad Sci* 1962; 24: 395-407.
8. Willard HF, Latt SA. Analysis of DNA replication in human X chromosomes using BrdU Hoecst techniques. *Am J Hum Genet* 1976; 28:213-27.

9. Kanda N, Yashida TH. Identification of the facultative heterochromatin X chromosome in females of 25 rodent species. *Cytogenet Cell Genet* 1979; 23:12-22.
10. Wolf SF, Mareni CE, Migeon BR. Isolation and characterization of cloned DNA sequences that hybridize to the human X chromosome. *Cell* 1980; 21:95-102.
11. Riggs AD. X inactivation, differentiation, and DNA methylation. *Cytogenet Cell Genet* 1975; 14: 9-25.
12. Lester SC, Korn NJ, DeMars R. Derepression of genes on the human inactiveX: evidence for differences in locus specific rates of derepression and rates of transfer of active and inactive genes after DNA mediated transformation. *Somat Cell Genet* 1982; 8.
13. Mohanda T, Sparkes RS, Shapiro LJ. Reactivation of an inactive human X chromosome; evidence for X inactivation by DNA methylation. *Science* 1981; 211: 393-96.
14. Wolf SF, Migeon BR. Studies of X chromosome DNA methylation in normal human cells. *Nature* 1982; 295: 667-71.
15. Wolf SF, Migeon BR. Implications for X chromosome regulation from studies of human X DNA. Cold Spring Harbor. *Symposium on Quantitative Biology* 1982; 47: in press.
16. Persico MG, Toniolo D, Nobile C, D'Urso M, Luzzatto L. cDNA sequences of human glucose 6-phosphate dehydrogenase cloned in pBR322. *Nature* 1981; 294: 778-80.
17. Kafatos FC, Jones CW, Efstratiatis A. Determination of nucleic acid sequence homologies and relative concentrations by dot hybridization procedures. *Nucleic Acids Res* 1979; 7: 1541-52.

THE USE OF MOLECULAR PROBES AND CHROMOSOMAL REARRANGEMENTS TO PARTITION THE MOUSE Y CHROMOSOME INTO FUNCTIONAL REGIONS

Eva M. Eicher
Sandra J. Phillips
Linda L. Washburn

Today the accepted hypothesis regarding mammalian sex determination is that a gene (or group of genes) on the Y chromosome initiates testicular development within the bipotential gonad and in the functional absence of this gene the gonad differentiates into ovarian tissue. This hypothesis is based on the findings that XO mice[1] and XO humans[2] are females and XXY humans are males.[3] Although most later findings support this hypothesis, we really know little about the genetic events responsible for mammalian sex determination.

The most direct approach for investigating the mechanisms of primary (gonadal) sex determination is to identify and characterize mutations that interfere with this process. Such mutations would cause development of ovarian tissue in an XY individual and testicular tissue in an XX individual. These mutations could be inherited as autosomal, X- or Y-linked genes, or they could involve chromosomal rearrangements. Of particular interest are structural rearrangements involving the Y chromosome, because correlation of the inheritance of such cytologically detectable alterations with phenotypic sex provides insight into where the testis-determining locus resides. In this paper we present genetic, cytological and molecular characterization of two inherited mouse Y chromosome rearrangements that have allowed us to partition the mouse Y chromosome into four functional units, one of which contains the testis-determining (*Tdy*) locus.

The first identified mouse mutation that interferes with primary sex determination was described by Cattanach *et al.*[4] Males that carry this mutation, named sex-reversed (gene symbol *Sxr*), produce four types of offspring in equal ratios when mated to normal females: XX females, XX males, XY *Sxr*-carrier males and XY non-*Sxr*-carrier males. That is, *Sxr*-carrier males transmit *Sxr* to half of their offspring - the XX males and XY *Sxr*-carrier males. Because a 1:1 ratio is exactly the ratio expected for inheritance of an autosomal dominant mutation, Cattanach *et al* assumed that *Sxr* was autosomally inherited.[5,6,7]

ISBN 0-12-492220-1

The most widely held hypothesis was that *Sxr* represented the translocation of a piece of Y chromosome, including the Y-linked testis-determining gene, *Tdy*, to an autosome. In addition, because XX *Sxr* males expressed the male-specific transplantation antigen H-Y (histocompatibility-Y),[8] the *H-Y* locus also was hypothesized to reside within the piece of Y chromosome involved in the *Sxr* mutation. Although a tremendous amount of time was spent in a coordinated effort by Cattanach's laboratory and our laboratory to locate the autosome carrying *Sxr* using classical mapping techniques, no linkage was found. Instead, the inheritance of *Sxr* was solved by molecular studies.

Singh *et al* showed that a satellite sequence isolated from the heterogametic female snake, the banded krait (*Bulgarus fasciatus*), hybridized to specific *Alu*I and *Hae*III restriction fragments in male but not female mouse DNA.[9] Hybridization *in situ* of this snake DNA sequence, designated Bkm, localized a concentration of Bkm-related sequences on the proximal (centromere) end of the normal Y chromosome.[10] When Singh and Jones hybridized Bkm DNA to *Alu*I restriction digested DNA from XX *Sxr* mice, the hybridization pattern was virtually identical to that observed in normal XY mice. These results were compatible with the idea that *Sxr* involved the translocation of a region of Y chromosomal DNA to another chromosome.[9,10] Hybridization *in situ* of Bkm-related DNA to chromosome preparations from XX *Sxr* and XY *Sxr*-carrier males showed that the distal end of one large chromosome contained a concentration of Bkm-related DNA. This chromosome was not seen in somatic cells of XX females or XY *Sxr*-carrier males. Singh and Jones also noted that the Y chromosome of XY *Sxr*-carrier males contained two distinct regions of Bkm-enriched sequences instead of one, as in normal males.[11]

Although Singh and Jones proposed that *Sxr*-carrier males contained two aberrant Y chromosomes, their cytological data did not support this explanation. (The two supposed aberrant Y chromosomes were two Y chromatids, an artifact of overexposure to colcemid.[6]) Eicher,[6] Hansmann[12] and Burgoyne[13] have independently suggested the following model to explain the molecular findings: The *Sxr* mutation originated from the translocation of a proximally located region from one Y chromatid to the distal end of the other Y chromatid (Figure 1). The proximal region involved in the translocation contains the testis-determining gene and the male-specific sequences that hybridize to Bkm DNA. Its duplicated position on the Y chromosome is located distal to the region of the Y chromosome that normally pairs with the X chromosome. An assumption underlying the model is that recombination normally occurs between the X and Y chromosomes in their pairing region, and that recombination also occurs between the abnormal Y chromosome and the X chromosome in *Sxr*-carrier males. Thus, all males that inherited the abnormal Y chromosome, hereafter designated Y^{Sxr}, produce four types of chromosomally distinct sperm: X^{Sxr}, Y, X and Y^{Sxr} (Figure 1). The X^{Sxr} chromosome is formed by the transfer of the distal region of the Y^{Sxr} chromosome to the

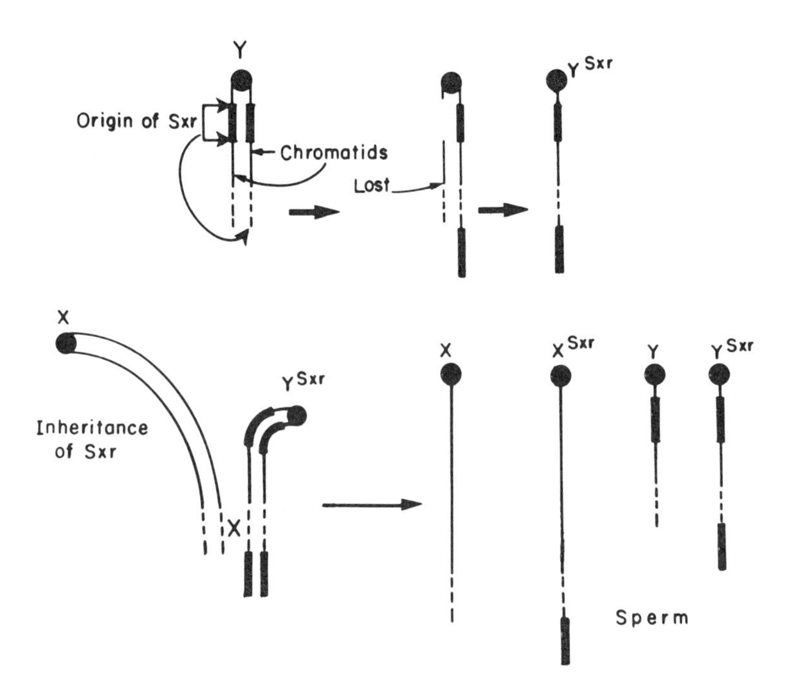

Fig. 1. Hypothesized rearrangement involving *Sxr*. A region of the Y chromosome located near the centromere is shown excised from one Y chromatid and inserted distally into the other Y chromatid. The chromosome containing the duplication is designated Y^{Sxr}. The formation of a chiasmata between the X-Y^{Sxr} bivalent results in four chromatids, thus sperm: X, X^{Sxr}, Y and Y^{Sxr}. (Figure modified after Eicher.[6])

X chromosome by recombination during meiosis. A normal Y chromosome is generated from the Y^{Sxr} chromosome after it has lost this distal region by recombination to the X chromosome. A normal X and the Y^{Sxr} chromosome are also inherited from XY^{Sxr} males because the recombination event involves only one of the two X chromatids and one of the two Y^{Sxr} chromatids (Figure 1).

Recently another Y chromosomal rearrangement has been found in the laboratory mouse.[6] This rearrangement arose in a male of the inbred strain LT/SvEi and was discovered in the process of transferring an X-linked isozyme variant of the phosphoglycerate kinase (*Pgk-1*) gene into the genome of the LT/SvEi inbred strain. From a trio mating involving two *Pgk-1^a*/*Pgk-1^b* females and an LT/SvEi (*Pgk-1^b*) male, several PGK-1AB males and PGK-1A females (phenotypes normally suggestive of XXY and XO karyotypes, respectively) were produced in addition to the expected progeny. Cytological and genetic analyses established that the LT/SvEi male carried an abnormal chromosome, hereafter designated Y*.

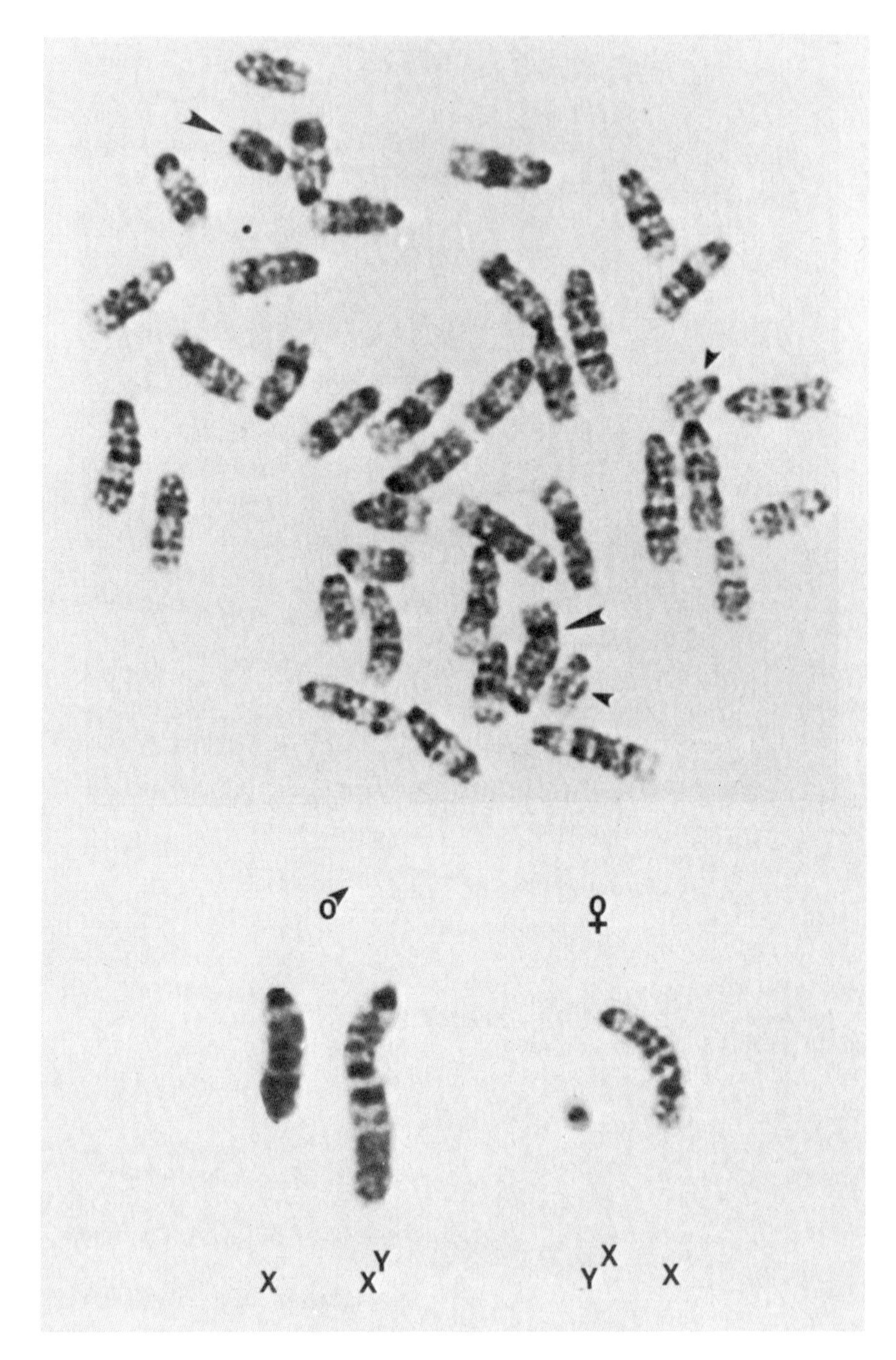
♂
♀
X
X^Y
Y^X
X

Further analyses of the inheritance of the Y* chromosome indicate that matings of XX females to XY* males result in five types of offspring: XY* and XX^Y males, and XX, XO and XY^X females. (Eicher and Washburn, unpublished) The X^Y chromosome consists of almost all of the X chromosome attached at its distal end to most of the Y chromosome and the Y^X chromosome consists of the proximal end of the Y chromosome, including its centromere, attached to the distal end of the X chromosome (Figure 2).

One hypothesis to explain the origin of the Y* chromosome is that it occurred by a transposition of all or part of its distally-located X-pairing region to a region just below its centromere (Figure 7). Another possibility is that the Y* chromosome resulted from the translocation of the distal region of the X chromosome to the Y chromosome just distal to its centromere (Figure 7). Regardless of which hypothesis is correct, as in the case of Y^{Sxr} chromosome, part of the Y* chromosome, including the *Tdy* gene, is transferred by recombination to the distal end of the X chromosome. In addition, because XX^Y males are H-Y positive and XY^X females are H-Y negative, the *H-Y* gene is also transferred from the Y* chromosome to the X chromosome during meiotic recombination.[14]

A major difference between the Y* and Y^{Sxr} chromosomes is that the segment of Y chromosome transferred by recombination from the Y* chromosome to the X chromosome is observable by light microscopy. (We have been unable to see any reproducible abnormalities associated with the distal end of an X^{Sxr} chromosome.) Another major difference between males carrying the X^{Sxr} or X^Y chromosome is their ability to produce functional sperm. The mating of an XO female to an XY^{Sxr} male produces $X^{Sxr}O$ male offspring that produce immotile sperm.[4] In contrast, the mating of an XO female to an XY* male results in X^YO males that produce motile sperm, often in sufficient numbers to sire offspring. (Eicher and Washburn, unpublished) This difference in fertility suggests that the region of the Y chromosome present in the X^Y chromosome but absent in the X^{Sxr} chromosome contains one or more loci involved with sperm motility.

The experiments we will describe involved the use of three different DNA probes to analyze the normal mouse Y chromosome and the abnormal Y^{Sxr} and Y* chromosomes together with their recombination products. These probes are of particular interest because the hybridization patterns observed with male versus female mouse DNA indicate that they recognize Y chromosomal DNA. The pur-

---

Fig. 2. G-banded mitotic metaphase plate prepared from an XY* male. Large arrow heads designate the X and Y* chromosomes, small arrow heads designate chromosome 19. Below the metaphase are shown the X and X^Y chromosomes from an XX^Y male and an X and Y^X chromosomes from an XY^X female. The X^Y and Y^X are reciprocal products produced by recombination of the X and Y* chromosomes.

pose of the experiments was to determine the location of specific Y chromosomal DNA sequences within the mouse Y chromosome and correlate these findings with our knowledge of the locations of the *Tdy* and *H-Y* loci. The probes used are described below:

1. M720 is a pBR322 clone of a xenotropic type C retrovirus endogenous to the murine species *Mus cookii*.[15] M720 hybridizes to four male-specific *Eco*RI fragments (14.5 kb, 11.8 kb, 7.5 kb and 4.4 kb) in inbred strains of mice. M720-related sequences are present throughout the inbred mouse genome, but are most abundant on the Y chromosome (~100 copies) and account for ~3% of Y chromosomal DNA.[15]

2. pBM19 is a pBR325 subclone containing a 1.8 kb *Eco*RI fragment from the mouse Y chromosome that hybridizes only to male DNA under the hybridization conditions used in these experiments (see below) (Phillips and Eicher, unpublished). This fragment was subcloned from a λ*Charon* 4A clone isolated from an *Eco*RI partial/*Eco*RI* partial library of C57BL/10 male DNA.[16]

3. The third probe, pErs5A, is a pBR322 subclone of satellite DNA from the female snake *Elaphe radiata*.[17] The 1.3 kb insert in pErs5A has been sequenced and contains 26 and 12 copies, respectively, of the tetranucleotides G-A-T-A and G-A-C-A. These simple repeats produce a sexually dimorphic hybridization pattern when pErs5A is hybridized against *Hae*III or *Alu*I digested male and female mouse DNA's. Hybridization *in situ* has shown that pErs5A-related sequences are distributed on all mouse chromosomes, but are concentrated on the mouse Y chromosome near its centromere (Epplen, personal communication), as is the case for the Bkm-related sequences in the mouse genome.[11]

Four μg of genomic DNA were digested to completion with the restriction enzymes *Eco*RI or *Alu*I according to the manufacturers instructions (Bethesda Research Laboratories, Inc.) and electrophoresed in 1% agarose gels buffered with 0.04M Tris, 0.02M NaOAC, 2mM EDTA, pH 7.4 with HOAC. The DNA was blotted onto nitrocellulose overnight using 10X SSC as described by Southern.[18] Filters were dried and baked for 2 hours at 80°C under vacuum. DNA probes were ^{32}P-labelled by nick-translation to a specific activity of >1x10^8 dpm/μg DNA.[19] Filters were prehybridized in 500 ml of 1X Denhardt's solution[20] and 3X SSC at 65°C for 2-5 hours. Hybridization was at 70°C for 16-20 hours in a hybridization solution (10 ml/filter) containing the nick translated probe (up to 5x10^7 dpm/^{32}P), 6X SSC, 0.05% SDS and competing DNA's and RNA's (poly [A], 8 μg/ml; poly [C], 8 μg/ml; yeast tRNA, 200μg/ml; *E. coli* DNA, 10 μg/ml; salmon sperm DNA,

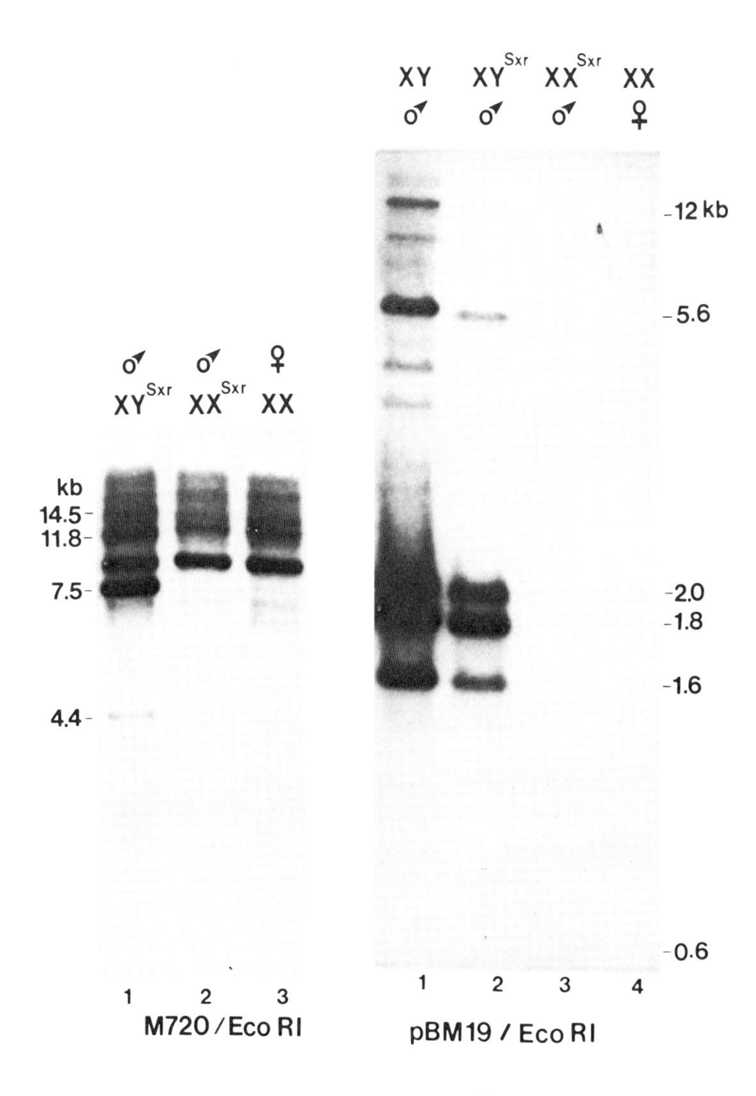

Fig. 3. Hybridization patterns relating to the Y^{Sxr} chromosome. *Eco*RI digested DNA's were probed with M720 (left panel) or pBM19 (right panel). DNA's used were XY^{Sxr} male, XX^{Sxr} male, XX female and XY male. The hybridization pattern obtained using DNA from an XY son (non-*Sxr* carrier) of an XY^{Sxr} male is similar to that observed in XY^{Sxr} DNA when probed with M720 (data not shown). Note that lane 2, right panel contained less DNA than present in the other lanes. Exposures were at room temperature, 24 hours for M720 and 36 hours for pBM19.

TABLE 1. Hybridization Patterns in DNA from Mice Carrying Abnormal Y Chromosomes or Their X-Recombination Derivatives

Y Chromosome	Chromosome Complement	Gonodal Sex	Male DNA Pattern: *EcoRI* M720 or pBM19	Male DNA Pattern: *AluI* pErs5A
Normal	XY	M	+	+
	XX	F	–	–
Y^{Sxr}	XY^{Sxr}	M	+	+
	XX^{Sxr}	M	–	+
	XY	M	+	+
	XX	F	–	–
Y*	XY*	M	+	+
	XX^{Y}	M	+	+
	XY^{X}	F	–	–
	XX	F	–	–

50 μg/ml). Filters were extensively washed at 52°C in 0.1X SSC and 0.05% SDS, air dried, and exposed at room temperature to Kodak XAR-5 X-ray film for lengths of time as indicated in the figure legends.

When M720 or pBM19 was used to probe *Eco*RI digested DNA's from XY^{Sxr} males and their XY (non-*Sxr* carrier) sibs, the pattern of hybridization was similar to that observed in normal males (Figure 3 and Table 1). However, XX^{Sxr} male DNA was not distinguishable from XX female DNA when probed with M720 or pBM19 (Figure 3 and Table 1). The absence of male-specific M720 or pBM19 hybridizing fragments in XX^{Sxr} male DNA indicates that all or most of the Y chromosomal region represented by these probes is not included in the Y chromosomal segment transferred by recombination from the Y^{Sxr} chromosome to the X chromosome. We also conclude that the part of the Y chromosome (>3%) represented by M720-like and pBM19 sequences is not closely linked to the *Tdy* gene.

In contrast, when these same DNA's (XY^{Sxr}, XY, XX^{Sxr} and XX) are digested with *AluI* and probed with pErs5A, a female *Elaphe radiata* satellite DNA, the observed patterns of hybridization are similar to those first reported by Singh *et al,*[9] who used satellite DNA isolated from a different species of female snake (*Bulgarus fasciatus,* banded krait) to analyze the *Sxr* mutation. As seen in Figure 4, the pattern of hybridization with XY^{Sxr}, XX^{Sxr} and XY DNA's was similar to that observed in normal male mouse DNA, whereas the pattern in the XX females was similar to that observed in other XX female mouse DNA (Table 1). These results indicate that most or all of the male-specific restriction fragments identified by the female snake satellite are present in the region of the Y^{Sxr} chromosome transferred by recombination to the X chromosome and that the Y chromosome of non-*Sxr* carrier males still retains copies of these sequences. These results support

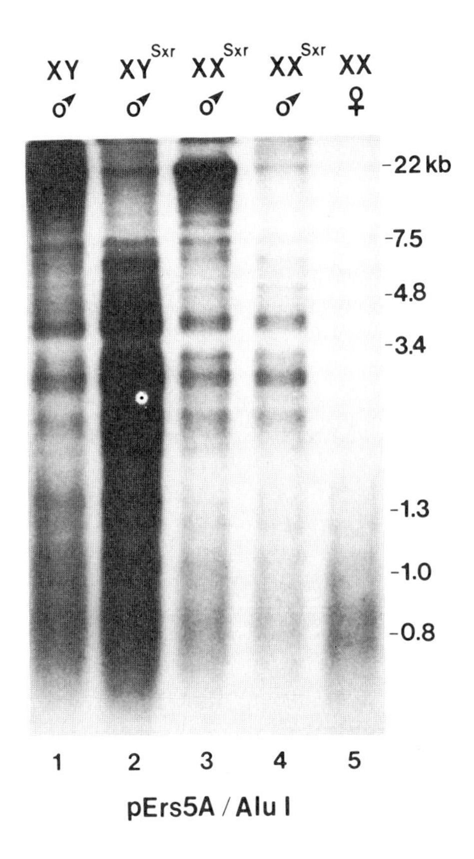

Fig. 4. Hybridization pattern relating to the Y^{Sxr} chromosome. *Alu*I digested DNA's were probed with pErs5A. DNA's, left to right, are: XY (non-*Sxr*-carrier), XY^{Sxr} Male, XX^{Sxr} Male, XX^{Sxr} Male, and XX Female. Exposure time was 36 hours at room temperature without intensifying screens.

the conclusions of Singh and Jones[11] that (1) *Sxr* involves a duplication of Y chromosomal sequences related to female snake satellite sequences and (2) the duplicated sequences reside on the Y^{Sxr} chromosome distal to the region that recombines with the X chromosome.

We analyzed DNA's obtained from mice carrying the Y* chromosome and its recombination products with the same probes used to analyze the Y^{Sxr} chromosome. The purpose of these experiments was to (1) gain insight into the nature of the rearrangement involving the Y* chromosome and (2) further our understanding of the organization of the normal Y chromosome. The results were that both M720 and pBM19 hybridized to *Eco*RI male-specific fragments in XY* and

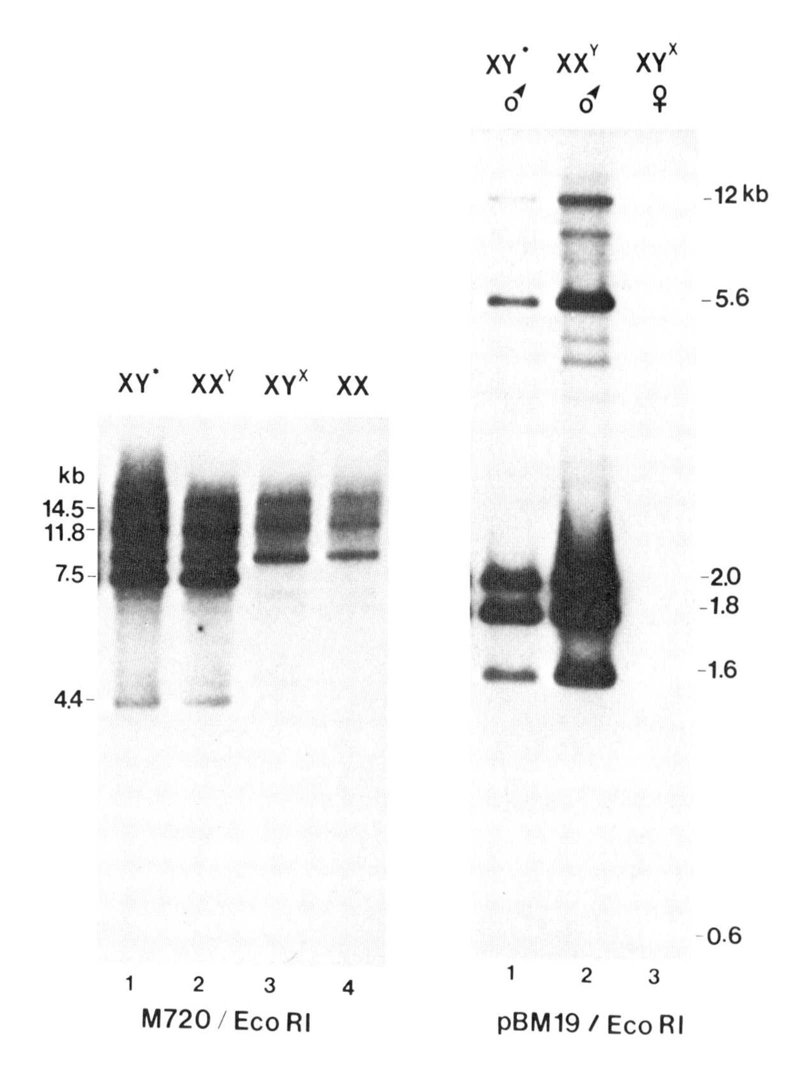

Fig. 5. Hybridization patterns relating to the Y* chromosome. *Eco*RI digested DNA's were probed with M720 (left panel) or pBM19 (right panel). The DNA's used were: XY* male, XX^{Y} male, XY^{X} female and XX female . Lane 1, panel 2 (XY*) contained less DNA than present in the other lanes. Exposure times 20 hours for M720 and 36 hours for pBM19.

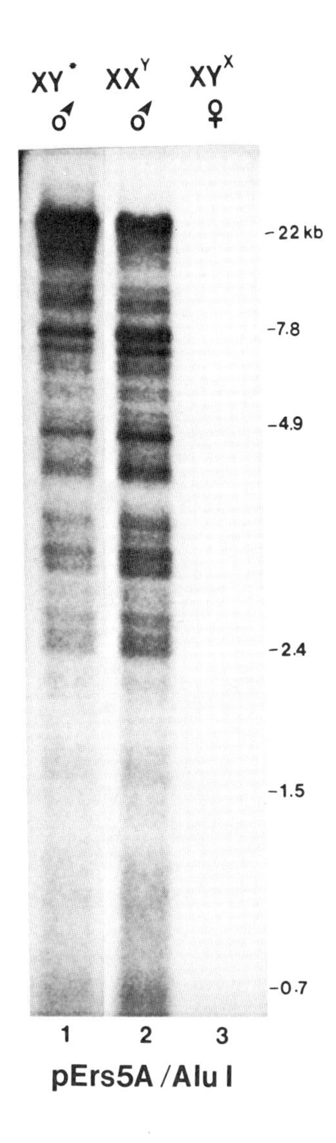

Fig. 6. Hybridization patterns related to the Y* chromosome. *Alu*I digested DNA's were probed with pErs5A. DNA's, left to right, are: XY* male, XX^Y male and XY^X female. Exposure time was 48 hours. Male specific *Alu*I fragments recognized by the pErs5A probe that are present in DNA from XX^Y males are not present in DNA from XY^X females (not shown).

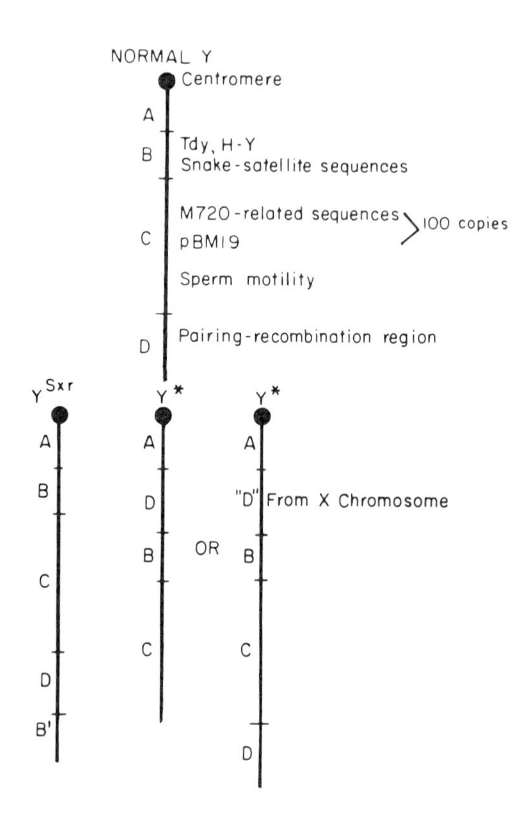

Fig. 7. The normal Y chromosome is shown divided into four functional regions: A contains the centromere; B contains the *Tdy* and *H-Y* loci as well as sequences homologous to female snake satellite DNA; C contains approximately 100 copies of sequences represented in these studies by the M720 and pBM19 probes respectively, and one or more genes involved with sperm motility; and D contains sequences homologous with those present on the distal end of the X chromosome such that it pairs with and recombines with the X chromosome. The organization of these same four regions in the Y^{Sxr} and Y* chromosomes are also shown.

XX^{Y} DNA but not to XY^{X} and XX DNA (Figure 5, Table 1). Similarly, a male-specific hybridization pattern was observed in *Alu*I digested XY* and XX^{Y} DNA probed with pErs5A but not in XY^{Y} and XX DNA (Figure 6, Table 1). These results indicate that all or most of the male-specific fragments recognized by M720, pBM19 and pErs5A are present in the region of Y chromosome transferred from the Y* chromosome to the X chromosome, X^{Y}, and absent from the resulting reciprocal recombination product, Y^{X}(Figure 2).

The results reported above combined with previous data allow us to partition the mouse Y chromosome into four functional regions (Figure 7):

Region A contains the centromere.

Region B includes the *Tdy* and *H-Y* loci, and the male-specific *Alu*I restriction fragments recognized by female snake satellite-related sequences. (This region is duplicated in the Y^{Sxr} chromosome and transferred by recombination to the X chromosome.) We place region B on the normal Y chromosome in a position just distal to region A because of the hybridization *in situ* studies of Singh and Jones.[10,11]

Region C contains all or most of the male specific *Eco*RI restriction fragments identified by the M720 and pBM19 probes and genetic information needed for sperm motility. Because we know that region C is not included in the piece of Y chromosome transferred from the Y^{Sxr} chromosome to the X chromosome but is included in the piece of Y chromosome transferred from the Y* chromosome to the X chromosome, we can position region C on the normal Y chromosome between regions B and D, where region D is the X-Y pairing region.[21,22]

Figure 7 shows the probable organization of these regions on the Y^{Sxr} chromosome. Note that region B is duplicated in the Y^{Sxr} chromosome, with the duplicated part, designated B′, located below region D. Two possible arrangements are shown for the Y* chromosome (Figure 7). In one arrangement, region D is located between A and B. In the other arrangement, region D is shown duplicated, with one region positioned normally (distal to region C) and the other, designated "D", located between regions A and B. This duplicated D region is designated "D" because it may be derived from the distal end of the X chromosome not from the Y chromosome *per se*. In either case, the Y* chromosome must contain an X-homologous pairing region in this area because recombination between the X and Y* chromosome has to occur just distal to region A to account for the G-banded patterns observed in the X^Y and Y^X chromosomes (Figure 2).

SUMMARY

Cytogenetic and molecular studies of two abnormal mouse Y chromosomes have enabled us to partition the normal Y chromosome into four functionally distinct regions. When ordered from centromere to telomere, these regions contain the centromere, the *Tdy* and *H-Y*[23] loci, genetic information involved with sperm motility, and an X-pairing-and-recombination segment. In addition, we have shown that the *AluI* male-specific fragments recognized by the female snake DNA probe pErs5A are located in the region containing the *Tdy* and *H-Y* loci and that the *Eco*RI male-specific fragments identified by the M720 and pBM19 probes are located in a region involved with sperm motility.

ACKNOWLEDGMENTS

This project was supported by a grant GM 20919 from the National Institutes of Health. The Jackson Laboratory is fully accredited by the American Association for Accreditation of Laboratory Animal Care. We are grateful to Joerg T. Epplen for sharing his pErs5A probe, and Steven Weaver for sharing his λ*Charon* 4A C57BL/10 male library. We thank Julie Drawbridge for help with chromosome preparations and Neal Copeland, Muriel Davisson and Nancy Jenkins for their valuable review of the manuscript.

REFERENCES

1. Welshons WJ, Russell LB. The Y chromosome as the bearer of male determining factors in the mouse. *Proc Natl Acad Sci USA* 1959; 45:560-66.
2. Ford CE, Jones KW, Polani PE, DeAlmeida JC, Briggs JH. A sex chromosome anomaly in a case of gonadal dysgenesis (Turner's Syndrome). *Lancet* 1959; i:711-13.
3. Jacobs PA, Strong JA. A case of human intersexuality having a possible XXY sex-determining mechanism. *Nature* 1959; 183:302-03.
4. Cattanach BM, Pollard CE, Hawkes SG. Sex-reversed mice: XX and XO males. *Cytogenetics* 1971; 10:318-37.
5. Cattanach BM. Sex reversal in the mouse and other mammals. *In:* Balls M, Wild AE, eds. The Early Development of Mammals. London:Cambridge University Press, 1975 :305-17.
6. Eicher EM. Primary sex determing genes. *In:* Amann RP, Seidel GE Jr, eds. Prospects for Sexing Mammalian Sperm. Boulder:Colorado Association University Press, 1982: 121-35.
7. Lyon MF, Cattanach BM, Charlton HM. Genes affecting sex differentiation in mammals. *In:* Austin CR, Edwards RG, eds. Mechanisms of Sex Differentiation in Animals and Man. New York:Academic Press, 1981:325-86.
8. Bennett D, *et al.* Serological evidence for H-Y antigen in *Sxr*, XX sex-reversed phenotypic males. *Nature* 1977; 265:255-57.
9. Singh L, Purdom IF, Jones KW. Conserved sex-chromosomes-associated nucleotide sequences in eukaryotes. *Cold Spring Harbor Symp Quant Biol* 1981; 45:805-14.
10. Jones KW, Singh L. Conserved repeated DNA sequences in vertebrate sex chromosomes. *Human Genet* 1981; 58:46-53.
11. Singh L, Jones KW. Sex reversal in the mouse (*Mus musculus*) is caused by a recurrent non-reciprocal crossover involving the X and an aberrant Y chromosome. *Cell* 1982; 28: 205-16.
12. Hansmann I. Sex reversal in the mouse. *Cell* 1982; 30:331-32.
13. Burgoyne PS. Genetic homology and crossing over in the X and Y chromosomes of mammals. *Hum Genet* 1982; 61:85-90.
14. Simpson E, Chandler P, Washburn LL, Bunker HP, Eicher EM. H-Y typing of karyotypically abnormal mice. *Differentiation* 1983, in press.
15. Phillips SJ, Birkenmeier EH, Callahan R, Eicher EM. Male and female mouse DNA's can be discriminated using retroviral probes. *Nature* 1982; 297:241-43.

16. Weaver S, Comer MB, Jahn C, Hutchison CA III, Edgell MH. The adult β-globin genes of the "single" type mouse C57BL. *Cell* 1981; 24:403-11.
17. Epplen JT, McCarrey JR, Sutou S, Ohno S. Base sequence of a cloned snake W-chromosome DNA fragment and identification of a male-specific putative mRNA in the mouse. *Proc Natl Acad Sci USA* 1982; 79:3798-3802.
18. Southern EM. Detection of specific sequences among DNA fragments separated by gel electrophoresis. *J Mol Biol* 1975; 98:503-17.
19. Maniatis T, Jeffrey A, Kleid DG. Nucleotide sequence of the rightward operator of phage λ. *Proc Natl Acad Sci* 1975;72:1184-88.
20. Denhardt DT. A membrane-filter technique for the detection of complementary DNA. *Biochem Biophys Res Commun* 1966; 23:641-46.
21. Schnedle W. End-to-end association of X and Y chromosomes in mouse meiosis. *Nature* 1972; 236:29-30.
22. Solari AG. The spatial relationships of the X and Y chromosomes during meiotic prophase in mouse spermatocytes. *Chromosoma (Berl)* 1970; 29:217-36.
23. Silvers WK, Gasser DL, Eicher EM. The H-Y antigen, serologically detectable male antigen and sex determination. *Cell* 1982; 28:439-40.

APPROACHES TO HUMAN GENETICS BASED ON DNA SEQUENCE POLYMORPHISM

Ray White
David Barker
Web Cavenee
Robin Leach
Dennis Drayna
Tom Holm
Jon Berkowitz
Mark Leppert

Several years ago we published a paper[1] arguing that it was possible in principle to construct a linkage map of the human using DNA sequence polymorphisms as genetic markers. It was further argued that such a map would provide an important tool for the study of human genetic disease. This paper reviews the progress of our laboratory in this endeavor and describes one of our first applications of this technology to the study of the genetics of human cancer.

The notion underlying the use of DNA sequence polymorphisms as genetic markers depends on two aspects of DNA technology. Restriction enzymes are used to sample DNA sequence polymorphism and unique loci are defined by single copy human DNA probes using the Southern method.

The first step is to define the polymorphic loci. In our first round of experiments with a small screen of long single copy DNA, we discovered a remarkable locus revealing at least ten alleles with *Eco*RI.[2] This locus has subsequently been mapped to chromosome 14[3] and more recently shown to be tightly linked to the *Gm* locus.[4] The molecular basis for this polymorphism was shown by digestion with other restriction enzymes to be an insertion/deletion series. Interestingly, subsequent experiments with a large number of probes have failed to reveal additional insertion/deletion polymorphism even though over 1,000 kb of DNA have been screened at arbitrary loci. It is noteworthy that other investigators have found insertion/deletion polymorphism at the insulin locus,[5] the Harvey-ras oncogene locus[6] and the *zeta* globin cluster (Chapman, personal communication).

We have, however, seen a number of polymorphisms apparently due to base pair substitution. In fact, nine of the ten polymorphic sites observed in this series, examining some 36 loci, were found using the enzymes *MspI* and *TaqI*. We in-

ISBN 0-12-492220-1

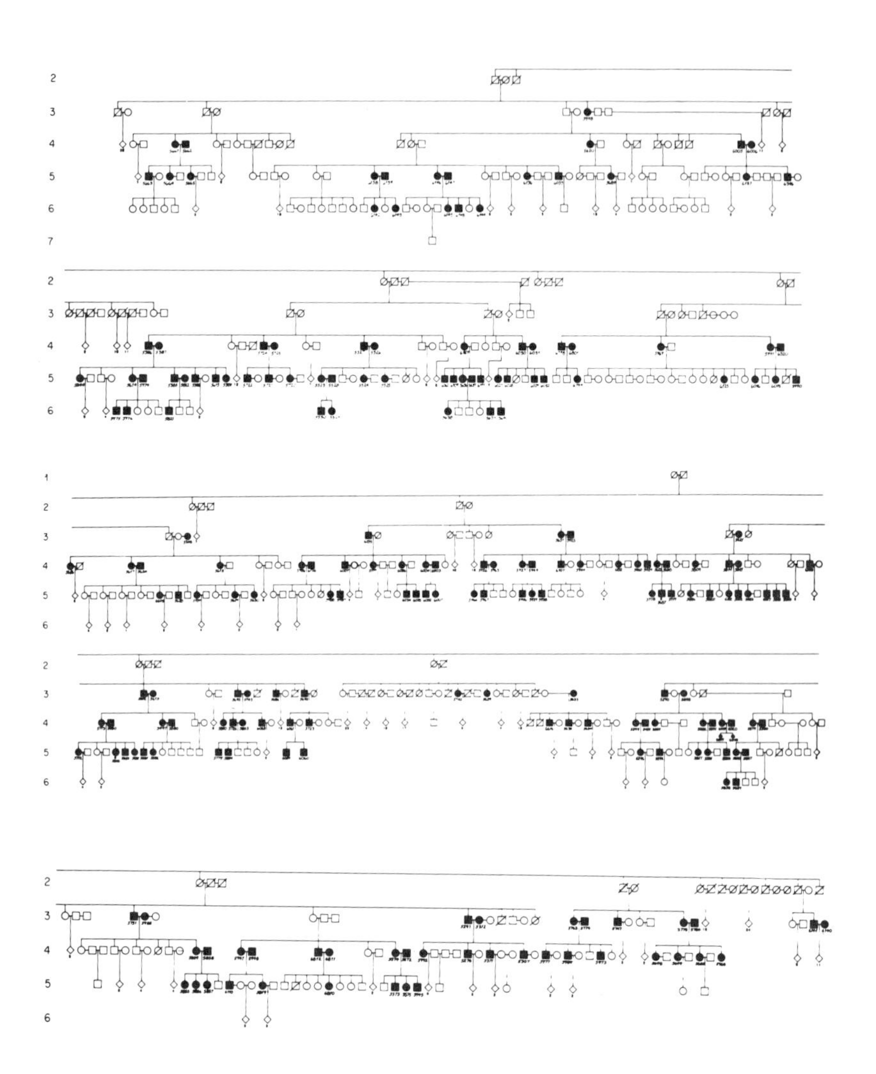

Fig. 1. Pedigree of Kindred 1085. Filled symbols indicate individuals who have been sampled for DNA.

terpret this result as due to a high relative mutation rate in the CG dimer sequence common to both enzymes. The estimated frequency of site polymorphism with these enzymes is supported by subsequent observations. For example, nine polymorphisms were observed with testing of 40 single copy sequences derived from a library believed to reflect sequences from chromosomes 13, 12 and 6p (Cavenee *et al,* in preparation) Polymorphic restriction fragment patterns from two probes localized to the X chromosome were also seen with the enzymes *MspI* and *TaqI*. One of these also showed homology to the Y chromosome, with a Y-specific *TaqI* allele.[7]

It should be noted that the usefulness of marker loci in linkage studies depends to some extent on the number and frequency of their alleles. For example, the average number of useful chromosomes (chromosomes whose recombination history can be determined) which can be expected to be observed in a set of 20 five-child sibships is approximately 50 for an interval defined by two two-allele marker loci, and approximately 100 for an interval defined by two four-allele loci. That is, the sample set for two-allele systems needs to be twice as large as for four-allele systems in order to obtain equivalent data. Not surprisingly, the expected LOD scores at any given recombination fraction are approximately proportional to the number of useful chromosomes. Furthermore, the efficiency, in terms of number of genotypic determinations required for equivalent data, even under reasonable sequential sampling protocols, is also significantly greater with four-allele systems; in fact, almost twice if the four alleles can be observed following digestion with a single restriction enzyme.

Fortunately, four multiallelic loci known to reside on chromosome 11p have provided an opportunity to actually begin gathering linkage data. The relevant portion of the pedigree we have started with is shown in Figure 1 and is being studied in collaboration with M. Skolnick and T. McClellan, who are gathering data on protein markers. The four marker loci are the *beta* globin cluster, the insulin gene locus, the Harvey-ras oncogene locus and an arbitrary locus, pADJ762, which shows polymorphism with *MspI, TaqI* and *BclI*. Examples of the *MspI* and *BclI* polymorphisms are shown in Figure 2.

Even with four such useful and informative loci, the yield from examination of some 120 individuals from Kindred 1085 was somewhat disappointing. Although the cumulative data indicate all four loci to be part of the same linkage group, the only linkage with a LOD score over three was for the ADJ762 - insulin interval at a recombination fraction 0.12. The analysis was useful, however, in pointing out the different requirements for a general linkage study as opposed to a genetic disease oriented mapping study. In retrospect it becomes obvious that, for a general linkage study, spouse parents are as important as parents within the family for providing phase information. Furthermore, the extended family or large kindred is also less critical for general linkage studies; and our current efforts, in addition to trying to complete 1085, are to ascertain large sibships with living grandparents.

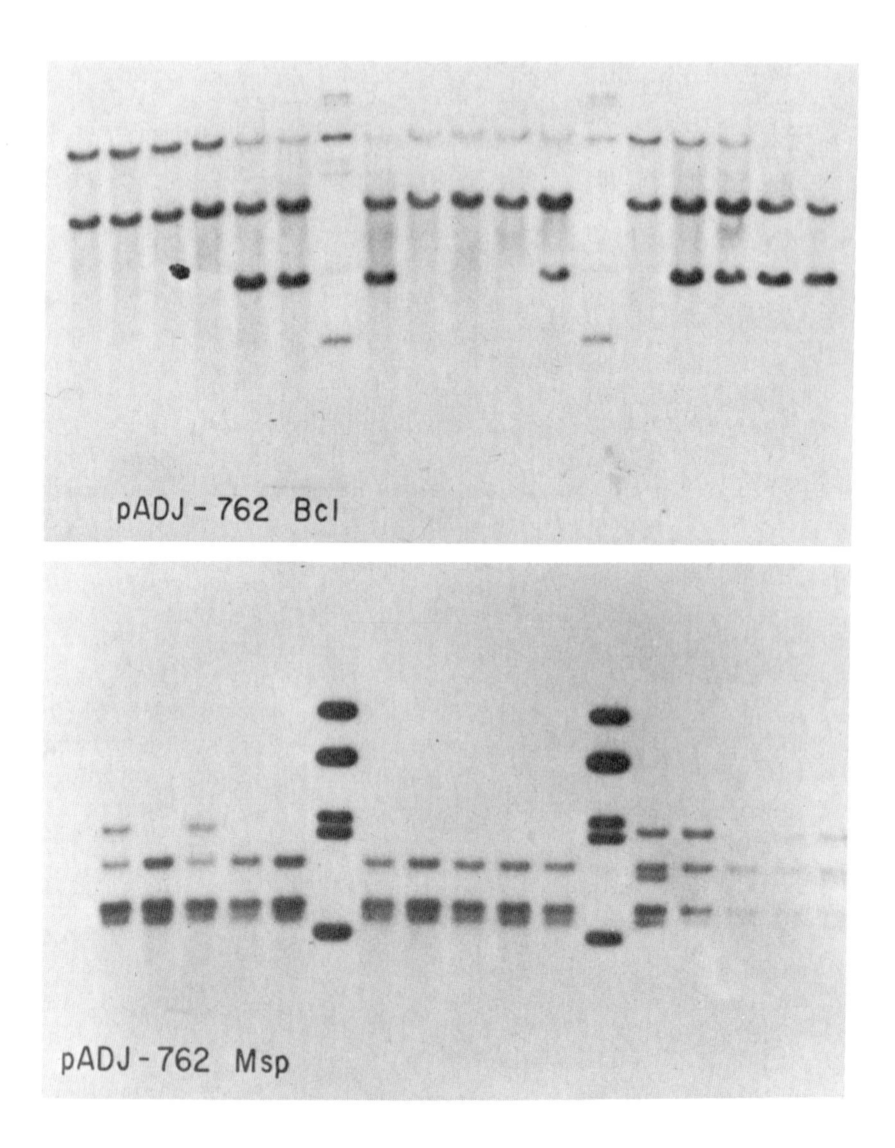

Fig. 2. Representative polymorphisms seen with the restriction enzymes *MspI* and *TaqI* at the locus defined by arbitrary probe pADJ762.

The recent observations that activated oncogenes can be isolated from human tumor cells has raised an intriguing question which can be addressed by genetic studies with DNA polymorphisms. Specifically, the hypothesis we would like to address is whether inherited mutations at oncogene loci can result in genetic predispositions to the development of specific tumors. For a first try at examining this hypothesis, we chose to examine a Kindred, 109, showing segregation for Gardner Syndrome, an autosomal dominant disorder whose phenotype includes a highly penetrant predisposition to colon cancer. Tissue samples from family members were provided by R. Burt and E. Gardner. Two oncogen loci have been examined for linkage with this disorder, the Kirsten-ras locus known to be activated in some human colon cancer cell lines and the Harvey-ras locus known to be activated in a human bladder cancer cell line.

The probe for the Kirsten-ras locus was provided by R. Weinberg[8] and an adjacent *TaqI* polymorphism was developed by us. Two families within the pedigree were informative at the marker locus as well as the Gardner locus. Multiple recombination events are required to explain the observed genotypes and it may be concluded that the Gardner locus and the Kirsten-ras locus are not tightly linked and, therefore, the disease syndrome does not result from a mutation at the Kirsten-ras locus.

The probe for the Harvey-ras locus was provided by M. Wigler and a polymorphism at the Harvey-ras locus has been described.[6] We have confirmed the Mendelian character of this polymorphism as revealed by the enzyme *TaqI*. The same two families within K109 were again informative and multiple recombinants are required to account for the observed data. Therefore, we must conclude that a mutation at this oncogene locus is also not responsible for the Gardner Syndrome.

Several conclusions may be drawn from these studies. First, we have shown that neither of two oncogene loci coincide with the Gardner locus; either mutation at other oncogene loci may be involved or some other locus not yet identified may be important in this inherited tumor predisposition. Second, we suggest that this approach to determining the involvement of candidate genes in specific inherited diseases is both efficient and effective; with a relatively modest effort, specific gene loci can be rigorously shown to be uninvolved with specific genetic diseases. Since we can anticipate that the number of cloned human genes will continue to rapidly increase, it will become important to determine their role in various disease states and the approach described here can make a contribution to this determination.

ACKNOWLEDGMENTS

Figures reprinted from the Banbury Report 14, Cold Spring Harbor Laboratory, 1983.

REFERENCES

1. Botstein D, White R, Skolnick M, Davis RW. Construction of a genetic linkage map in using restriction fragment length polymorphisms. *Gene* 1980; 2:95-113.
2. Wyman AR, White R. A highly polymorphic locus in human DNA. *Proc Natl Acad Sci USA* 1980; 77:6754-58.
3. de Martinville B, Wyman AR, White R, Francke U. Assignment of the first random restriction fragment length polymorphism (RFLP) locus (D14S1) to a region of human chromosome 14. *Amer J Hum Gen* 1982; 34:216-26.
4. Balazs I, Purrelo M, Rubinstein P, Alhadeff B, Siniscalso M. Highly polymorphic DNA site D14S1 maps to the region of Burkitt lymphoma translocation and is closely linked to the heavy chain *Gamma* 1 immunoglobulin locus. *Proc Natl Acad Sci USA* 1982; 79:7395-99.
5. Bell GI, Selby MJ, Rutter WJ. The highly polymorphic region near the human insulin gene is composed of simple tandemly repeating sequences. *Nature* 1982; 295:31-4.
6. Goldfarb M, Shimizu K, Perucho M, Wigler M. Isolation and preliminary characterization of a human transforming gene from T24 bladder carcinoma cells. *Nature* 1982; 296: 404-14.
7. Page D, *et al.* Single-copy sequence hybridizes to polymorphic and homologous loci on human X and Y chromosomes. *Proc Natl Acad Sci USA* 1982; 79:5352-56.
8. Murray MJ, Shilo B, Shih C, Cowing D, Hsu H, Weinberg RA. Three different human tumor cell lines contain different oncogenes. *Cell* 1981; 25:355-61.

HIGH RESOLUTION CHROMOSOME MAPPING OF CLONED GENES AND DNA POLYMORPHISMS

Thomas B. Shows
Bernhard U. Zabel
James V. Tricoli

INTRODUCTION

A precise human gene map is an essential prerequisite for understanding the molecular genetics of human biology and disease. There are 24 different chromosomes and, thus, 24 nuclear gene maps. There is also a 25th map comprising mitochondrial DNA. Eventually all these maps will be completely deciphered, which will be of inestimable value to human biology and molecular disease. It is important to map as many human genetic markers as possible, to assign them to a specific site on a chromosome, and ultimately to a specific nucleotide sequence on a chromosome.

The advent of recombinant DNA technology has given the human geneticist a new set of molecular markers that theoretically spans the entire genome.[1] These markers are the genes themselves, not their products, as well as undefined DNA segments that have been cloned; they represent very large numbers of markers for genetic studies. In addition to providing this enormous and profitable set of new markers, molecular technology has provided ways for high resolution gene mapping. In these initial phases of study, the degree of resolution is not as fine as ultimately desired. However, the technology centering around *in situ* hybridization (see below) is capable of identifying gene sites in bands and sub-bands of stained human chromosomes observed at the light microscopic level.[2] Specific-site chromosome assignment depends on having cloned probes without repetitive sequences of specific genes and undefined DNA segments. The radiolabelled probes can be hybridized to homologous sequences on metaphase chromosomes; after exposure to autoradiographic emulsion, silver grains identify the chromosomal site. Also, mapping of genes and gene products to chromosomal bands and sub-bands is made possible by the use of human translocations in human-mouse cell hybrids.[3] After several different segments of a human chromosome have been tested in independent

ISBN 0-12-492220-1

cell hybrids, the smallest overlapping segment identifies the location of the gene. At the current level of high resolution mapping, there are principally two types of strategies which are presented below.

Now that genes and gene products can be located to specific sites on human chromosomes, it is necessary to map enough DNA polymorphisms along each chromosome to carry out family studies for gene linkages and the linear ordering and genetic organization of known genes and diseases. The recombinant DNA technology has made it possible to detect DNA polymorphisms at a level that will provide enormous numbers of new polymorphic markers.[4] These polymorphisms represent DNA fragment-size polymorphisms, which can be mapped by *in situ* hybridization or by the cell hybrid strategy.[1,2]

The chromosome assignment of cloned genes and DNA polymorphisms, and the high resolution mapping of most of these markers is presented here. These molecular markers were chosen to study human disease. Most of the markers are either associated with a specific disease and/or are DNA polymorphisms that can be used as markers to study syntenic human disorders. Thus the routine high resolution mapping of inherited disease is possible. This information will considerably strengthen the human gene map and allow the human geneticist to predict, counsel, treat, and possibly correct human molecular disease.

SMALLEST CHROMOSOMAL SEGMENT

Of the few techniques available for assigning a human gene to the smallest chromosomal segment (SCS), the methods of rearranged chromosomes segregating in cell hybrids and the transfer of genes on a defined chromosomal segment into recipient cells have yielded the most information.[3] In human-rodent cell hybridization studies, if parental human cells have a chromosomal translocation or deletion, then human genes can be assigned to specific chromosomal segments that are retained in cell hybrids. For determining the smallest chromosomal segment, a series of independent cell hybrids with different lengths of a specific chromosome are tested for the marker, and the smallest segment is determined by the length of chromosome that overlaps all the different chromosomal segments. This procedure has been discussed, diagramed, and used extensively to identify the SCS.[3,5,7] The most documented fine resolution mapping of human genes by the SCS methodology are *PGM1*(phosphoglucomutase) localized to the p22.1 site[7] and the X-linked *G6PD* gene localized to the p28 site.[3,8] Because sufficiently large numbers of chromosomal rearrangements are not known for many chromosomes, a host of other genes have been mapped only to larger segments.[9] This method of mapping to the smallest segment is thus limited to the availability of large quantities of rearranged human chromosomes.

Using microcell-mediated gene transfer, chromosome-mediated gene transfer, gene transfer with defined segments of chromosomes and DNA, and vector-mediated gene transfer, the SCS of a few genes has been determined.[3] These systems have relied on using selectable markers and markers on a region of a chromosome that has been translocated to a selectable marker chromosome. Currently, these methodologies are limited to the few selectable markers known and their closely syntenic markers.

We have employed a number of cloned genes and DNA segments important in studying human disease or as genetic markers for mapping disease in order to determine high resolution mapping by cell hybridization strategies (described above) and by *in situ* hybridization site-specific mapping (described below). Of the cloned genes we have studied, a cloned amylase (*AMY*) gene (most likely recognizing both *AMY1* and *AMY2*[10]) offers an excellent example for comparing both fine resolution mapping determined by SCS mapping and *in situ* hybridization site-specific mapping. Combining cell hybrid mapping and gene linkage strategies, Cook and Hamerton[7] deduced the SCS encoding the *AMY* genes to be the p22.1→q11 region on chromosome 1. We have utilized cell hybrids retaining the pter→p21 chromosome 1 segment and combined recombinant DNA technology and cell hybrid strategies[1] to map the amylase gene.[10] It was found that cell hybrids with this 1pter→p21 segment were AMY positive, and thus localized the *AMY* genes to the 1p22.1→p21 segment by the smallest overlapping segment method diagramed in Figure 1, left.[10] Physical (cell hybrid) and Mendelian (gene linkage) mapping have been combined to determine the SCS. This high resolution mapping corresponds exactly with the results of *in situ* hybridization (see below). Although the SCS mapping strategy marks a significant advance in high resolution gene mapping, the SCS detectable by current methods of microscopy remain large and encode several hundred genes.

IN SITU HYBRIDIZATION

The importance of this technique for fine resolution mapping has been discussed.[1] Recent advances in *in situ* hybridization techniques have made it possible to map single-copy DNA sequences to specific chromosomal sites directly on fixed human metaphase chromosomes.[11-13] Figure 2 diagrams the procedure of *in situ* hybridization that has been refined and optimized[2] to assign the several cloned genes and DNA polymorphisms presented here to specific chromosomal sites. This procedure increases sensitivity and employs high resolution chromosomal banding techniques to optimize the specific chromosomal location. Features that have increased the resolving power include using large prometaphase chromosomes from synchronized lymphocyte cultures, using high specific activity probes, using dextran sulphate to promote formation of probe networks, and high resolution chromosome banding techniques through the emulsion.

Utilizing the procedure in Figure 2, the same amylase probe employed for SCS mapping (Fig. 1, left) was employed for *in situ* hybridization. *AMY* was con-

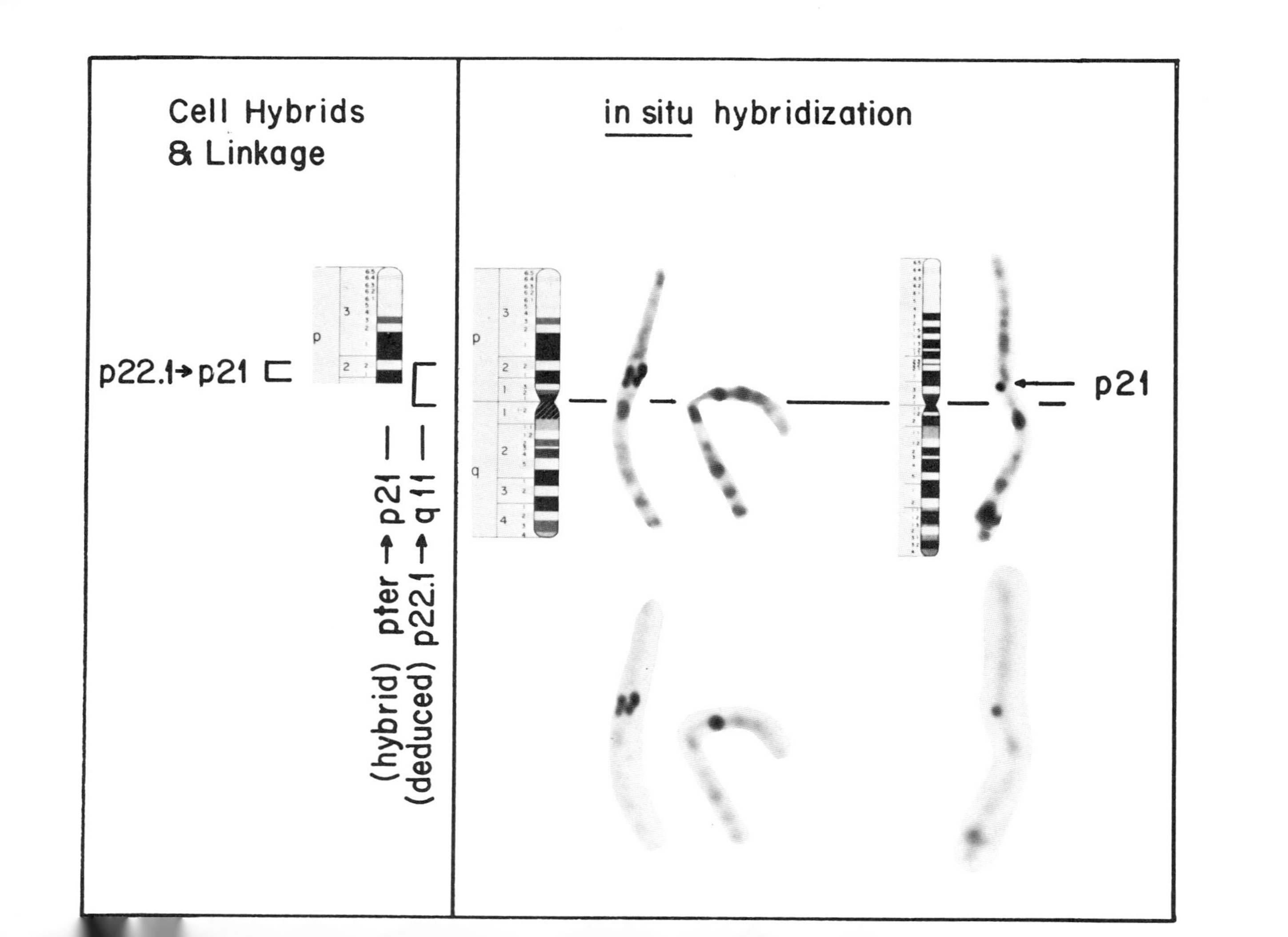
Cell Hybrids
& Linkage
p22.1→p21
(hybrid) pter → p21
(deduced) p22.1 → q11
in situ hybridization
p21

vincingly localized to the p21 site on chromosome 1 (Fig. 1, right).[2] The upper chromosomes in the figure illustrate the high resolution banding and the 1p21 site by *in situ* hybridization; the lower chromosomes are prepared with a different filter, and best demonstrate the location of silver grains formed as the result of β-emission from the radiolabelled probe.

Both methods of site-specific mapping of *AMY* (by smallest chromosome segment mapping and by *in situ* hybridization) correspond completely and were obtained independently using the same *AMY* probe. These results illustrate the corresponding mapping results of two high resolution mapping techniques, but also emphasize the direct nature of *in situ* hybridization for mapping the gene itself to a specific chromosomal site.

Utilizing these two procedures for site-specific chromosomal mapping, cloned genes and DNA restriction polymorphisms either involved in molecular disease or as markers for gene linkage studies have been chromosomally assigned and are discussed below.

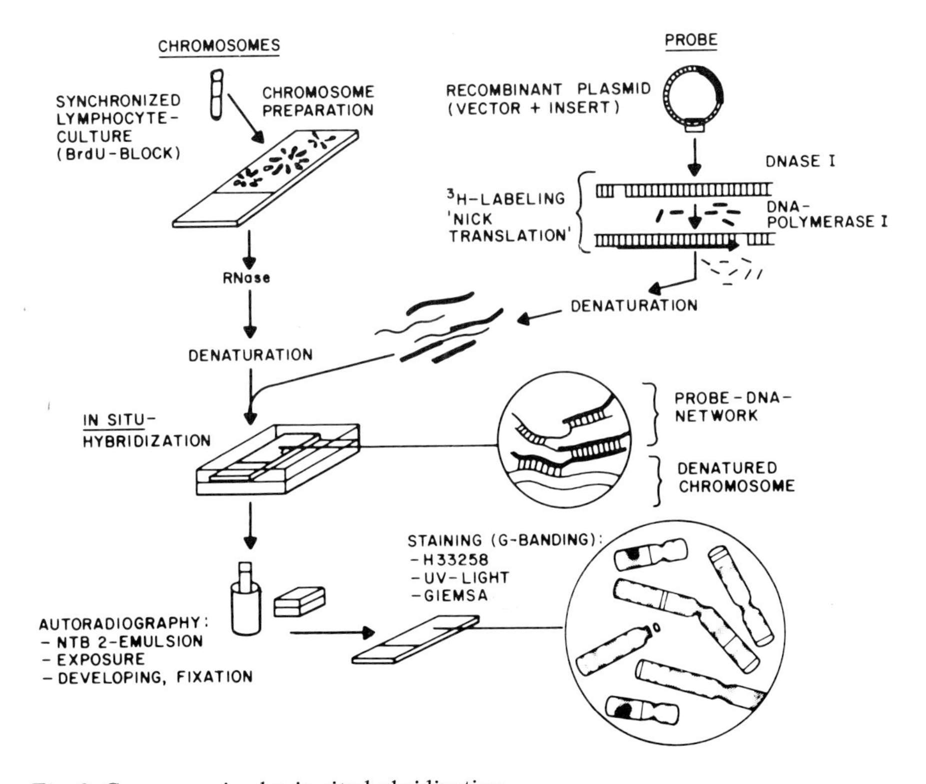

Fig. 2. Gene mapping by *in situ* hybridization.

Figure 1. Chromosome 1 mapping of amylase.

HIGH RESOLUTION MAPPING OF CLONED HORMONE GENES

There are a sizable number of hormone structural genes that have been cloned. When these genes are altered, abnormal hormone activity results in a multitude of disorders. Mapping these genes will aid in genetically characterizing the associated diseases, add mapped markers to the human genome, and serve as chromosomal sites around which it may be possible to uncover a DNA restriction polymorphism either within the gene or nearby.

The human genes we have mapped are tabulated in Table 1A. They include: insulin (*INS*), growth hormone (*GH*), chorionic somatomammotropin (*CSH*), prolactin (*PRL*), parathyroid hormone (*PTH*), proopiomelanocortin (*POMC*), and somatostatin (*SST*). The structural gene for insulin, an important polypeptide in carbohydrate metabolism with mutants involved in certain types of diabetes, has been located on the short arm of chromosome 11.[14,15] Using *in situ* hybridization, Zabel *et al.*[2,16] have located the gene at the p15→p14 site on chromosome 11. *GH, CSH,* and *PRL* are genes which share structural homology at both the amino acid

TABLE 1. Mapping Cloned Genes

Chromosome	Marker	Region	*in situ*	DNA size polymorphism	Refs.
A. Hormone Genes					
2	POMC	p23	+		2,16,21
3	SST	q28	+	+	2,16,22
6	PRL	pter→p23		+	19,40
11	INS	p15→p14	+	+	2,14-16,36
11	PTH	p11→pter			20
17	CSH	q22→q24	+		17,18
17	GH	q22→q24	+		17,18
B. Proto-Oncogenes					
12	KRAS2			+	26
20	SRC				25
C. Interferon Genes					
9	IFA	pter→q12			31,32
9	IFB	pter→q12			31,32
12	IFG	p12→qter			33
D. Enzyme Genes					
1	AMY	p21	+		2,10
16	CTRB			+	34
E. tRNA Genes					
6	TRM1	p23→q12			35
6	TRM2	p23→q12			35

and nucleotide level. GH and CSH (also called placenta lactogen) are produced in the pituitary and placenta, respectively, and bear an 85% amino acid sequence homology; PRL, produced in the pituitary, bears a 26% homology with GH and a 27% homology with CSH. The genes *GH* and *CSH* have been located on chromosome 17[17] and the 17q22→q24 region by *in situ* hybridization.[18] Prolactin is not on 17, but on chromosome 6.[19]

The parathyroid hormone gene has been located on the short arm of chromosome 11 (pter→p11) along with *INS*, although there is no indication as yet of how close they are on the physical map.[20] *POMC* is located on the short arm of chromosome 2[21] and specifically at the p23→p22 site.[16] *POMC* is a precursor protein produced in the pituitary that is processed into the important smaller peptide hormones of adrenocorticotropin (ACTH), lipotropin (LPH), α-melanocyte stimulating hormone, a corticotropin-like intermediate lobe protein, β-melanocyte stimulating hormone, and β-endorphin.[21] SST is a hormone and possible neurotransmitter that inhibits the secretion of several hormones and regulates gastrointestinal function. The *SST* gene has been located on the long arm of chromosome 3 (q21→qter) using cell hybrids[22] and to the q28 site by *in situ* hybridization.[2,16]

The importance and interactions of hormones would suggest there might be a coordinated regulation and organization of expression. Yet the genes we have mapped do not indicate there has been any concerted pressure for hormone structural genes to be clustered on the same chromosome. The *GH, CSH, PRL* family indicates that in evolutionary time, related and linked genes became separated.

MAPPING HUMAN PROTO–ONCOGENES

Evidence indicates that several normal genes with essential roles in cellular biochemistry are the progenitors for oncogenes of acutely transforming retroviruses of mammals and hybrids.[23] Several of these proto-oncogenes, in a wide diversity of species, have been implicated in malignant transformation. Evidence is building that normal genes in special situations may acquire oncongenicity. It is of great interest to determine where these proto-oncogenes reside on human chromosomes, if they are located on chromosomes associated with specific cancers, and at sites of structural rearrangements on these chromosomes. We have utilized cloned retroviral oncogenes to detect homologous human proto-oncogene sequences in interspecific cell hybrids by Southern filter hybridization.[24] Human proto-oncogenes were identified and human chromosome assignments were determined.[25-27] Using the cell hybrid-recombinant DNA strategy for mapping human proto-oncogenes, we have mapped two (Table 1B).[27]

A human cellular sequence homologous to the Rous sarcoma viral *src* gene was mapped to human chromosome 20.[25] Fine mapping of this gene has yet to be determined. The viral *src* gene is responsible for malignant transformation in cells. It has not yet been clearly defined if chromosome 20 has been implicated in a specific cancer. It has been tentatively suggested that an abnormal chromosome 20 is implicated in multiple endocrine neoplasia.[28]

The second human proto-oncogene mapped was the cellular homolog of the oncogene of Kirsten murine sarcoma virus.[26] This probe was isolated from SW480 human colonic adenocarcinoma cells using DNA-mediated transfection and molecular cloning techniques.[29] The human gene designated KRAS2 was assigned to chromosome 12 using cell hybrids and Southern filter hybridization techniques.[26] Currently there is little hard evidence that rearrangements of chromosome 12 are involved in any specific cancer, although a recent report has implicated the long arm of chromosome 12 in lymphoproliferative disease.[30]

Since several different human parental cells were employed in the cell hybrids tested, the chromosome assignments listed above (Table 1B) represent constant and native chromosomal locations of human proto-oncogenes. While *src* and KRAS2 were not apparently associated with chromosomal rearrangements, another human proto-oncogene, *MYC*, has been assigned to human chromosome 8 by Sakaguchi *et al.*[27] This chromosome is involved in specific translocations (8;14, 2;8, or 8;22) associated with Burkitt's lymphoma cell lines.[27]

MAPPING INTERFERON GENES

The interferons are a family of proteins that convey viral resistance and affect cell proliferation and the immune response.[31-33] There are three distinct forms of human interferon: the leukocyte form (IFN-α), the fibroblast form (IFN-β), and the immune form (IFN-γ). DNA cloning has identified more than 12 genes coding for IFN-α, and single genes for IFN-β and IFN-γ. Nucleotide sequencing shows that the genes for IFN-α, designated *IFA*, are closely related, with 80-95% nucleotide sequence homology. The gene for IFN-β, designated *IFB*, contains 40-50% sequence homology to *IFA* genes. Introns are not contained in the *IFA* and *IFB* genes, but the IFN-γ gene, designated *IFG*, has three introns and little DNA sequence homology to *IFA* and *IFB*. These observations suggest the genes for IFN-α and IFN-β are closely related evolutionarily.

Gene mapping was accomplished by filter hybridizations of cloned interferon cDNA to DNA from human-mouse cell hybrids (Table 1C). Genes for IFN-α and IFN-β were mapped to the pter→q12 region of chromosome 9 using a 9;17 translocation.[31,32] The gene for IFN-γ was mapped to the p12→qter region of chromosome 12.[33] Based on mapping of gene families, it is not surprising that the closely related *IFA* and *IFB* genes are located and clustered on chromosome 9, while the unrelated *IFG* gene is located on a different chromosome (chromosome 12). A similar observation was made (see above) for the *GH, CSH, PRL* hormone gene family.

MAPPING ENZYME GENES

We have mapped two cloned genes coding for enzymes (Table 1D). The amylase genes were mapped to the p22.1→p21 segment of chromosome 1 by cell hybrid strategy and to the smaller p21 segment by *in situ* hybridization (Fig. 1).

In addition, a probe for the chymotrypsinogen B (*CTRB*) gene was employed in cell hybrid mapping studies. *CTRB* is a member of the serine protease family and has been mapped to chromosome 16.[34] The same chromosome contains the gene for haptoglobin, a hemoglobin binding protein which is related to serine proteases. High resolution mapping for *CTRB* is in progress.

MAPPING METHIONINE TRANSFER RNA GENES

Molecular evidence suggests that there are at least 12 methionine initiator transfer RNA genes ($tRNA^{met_i}$).[35] However, the $tRNA^{met_i}$ genes are not directly adjacent based on recombinant DNA studies. In other organisms, tRNA genes are either scattered over the genome, clustered, or arranged in a tandem repeat. Using cloned $tRNA^{met_i}$ genes, the mapping of these genes was studied in cell hybrids to determine their organization in humans (Table 1E). Two clones containing different human methionine transfer RNA genes (*TRM1, TRM2*) were employed. These clones were known to be nonallelic based on restriction maps of the tRNA gene flanking sequences and the banding patterns seen when these clones were hybridized to digested genomic DNA.

Using human-mouse cell hybrids and Southern filter hybridization, the two genes were found to be assigned to the same human chromosome 6. Using cell hybrids made from human cells containing translocations of chromosome 6, both genes were localized in the p23→q12 segment of chromosome 6.[35] These data suggest that although these two $tRNA^{met_i}$ genes are not within 20 kb or so, they are located together and also suggest that other $tRNA^{met_i}$ genes may reside in this region. Further high resolution mapping will determine whether they are located on either the p or q arms or both.

DNA RESTRICTION SITE POLYMORPHISMS

DNA polymorphisms can arise from single-base pair substitutions that create or destroy a restriction enzyme recognition site, or from additions or deletions of DNA, or from other rearrangements – such as inversions. These changes can give rise to fragment-length variations that can be visualized by Southern filter hybridization.[36,37] Such DNA fragment-size variants should segregate in a codominant Mendelian fashion, so that an individual could be homozygous or heterozygous depending upon the variant carried by each chromosome homolog.[1] Based on the number of sequence polymorphisms detected in the β-globin cluster of man, it has been estimated that the human haploid genome would contain $3x10^7$ sequence variants.[38] This translates to about one every 100 base pairs. Since there are 24 human chromosomes per haploid genome, there could be on the average of about 10^6 sequence polymorphisms per chromosome.

Even if only 10% of such sequence polymorphisms can be recognized by restriction enzymes, there would be about 10^5 possible polymorphisms per chromosome that could be used as markers. If the average chromosome contains 10^8 base pairs, a sequence polymorphism would occur for each kilobase of DNA. Therefore, a good chance exists that a single-copy cloned probe will identify a DNA polymorphism. Thus, DNA polymorphisms represent a significant source of new markers for human genetic studies.[4] If DNA polymorphisms were mapped at equal distances on every human chromosome, then any inherited disease could be tested in family studies with a panel of DNA polymorphisms; close linkage would determine the chromosomal location and possible chromosomal site. Using mapped DNA polymorphisms would be more efficient in linkage tests than a panel of unassigned DNA polymorphisms. It has been estimated that a panel of 400-500 DNA polymorphisms spread about 20 cm apart would provide the necessary panel.[39] A new era in linkage and mapping is possible using mapped DNA polymorphisms. This strategy requires high resolution mapping of these markers to be effective. Our efforts have been to chromosomally map DNA polymorphisms recognized by single-copy cloned genes and undefined DNA segments (Table 2).

TABLE 2. Mapping DNA Restriction Polymorphisms

Chromosome	Marker	Region	*in situ*	Refs.
1	AMY	p21	+	10
3	D3S1*	q12	+	2,16,43
	12-32*	q12→qter		41
	SST	q28	+	2,16,22
5	12-65*	pter→q35		41
6	PRL	pter→p23		40
9	12-8*	pter→q12		41
11	INS	p15→p14	+	2,16,36
	ADJ-762*	pter→p11		41
12	KRAS2			26
16	CTRB			34
17	12-2*			41
18	12-62*			41
21	237D*			44
22	MS3-18*	pter→q13		41

* DNA segment

DNA Polymorphisms Recognized by Cloned Genes

DNA from 10–20 individuals is examined for DNA polymorphisms with each cloned gene that we map (Table 1). With only a limited number of restriction enzymes tested, a small population sample, and not all cloned genes as yet tested, we have detected DNA polymorphisms in all but two cloned genes examined.[22,26,34,40] Three of these polymorphisms have been mapped by *in situ* hybridization (Tables 1 & 3).

TABLE 3. *In Situ* Site Specific Mapping[2,16]

Chromosome	Marker	*In situ* Hybridization	Polymorphic	Somatic Cell Hybridization
1	AMY	p21		p22.1→p21
2	POMC	p23		pcen→pter
3	D3S1	q12	+	p21→q21
	SST	q28	+	q21→qter
11	INS	p15→p14	+	pter→p13

DNA Polymorphisms Recognized by Undefined DNA Segments

Cloned undefined DNA single-copy segments of the human genome represent an immense number of new genetic markers.[1] Based on the above discussion, a significant number of these undefined segments will recognize DNA polymorphisms. We have obtained several of these undefined DNA polymorphisms from collaborators (see Table 2 refs.) and have mapped them; their chromosomal location is in various stages of high resolution mapping.[41] Table 2 lists those undefined DNA markers (referred to as DNA segments in Table 2) that recognize DNA polymorphisms, and their chromosome assignment and subregional localization (if determined). Nine new polymorphic undefined DNA markers have been mapped.

A total of 15 new DNA polymorphisms have been mapped in a relatively short period of time, which adds strength to the hope that a large panel of these new markers will soon be available for large scale family studies and the mapping of disease.

GENES MAPPED BY *IN SITU* HYBRIDIZATION

Currently the most direct way to map a gene to a specific site on a chromosome is *in situ* hybridization (Figs. 1 & 2). The resolution of this procedure makes it possible to localize a gene to a specific chromosome band (stained regions of a chromosome) and sometimes to a sub-band after high resolution banding techniques. This level of resolution still spans about $5x10^3 - 10^4$ kb but makes a significant advance in site-specific gene mapping. This procedure is limited to those single-copy genes and undefined DNA segments that have been cloned.

A more time-consuming high resolution procedure is the smallest chromosomal segment determination using cell hybrids. The serious limitation of this procedure is finding, often by chance, a chromosome rearrangement with a breakpoint that reduces the chromosome segment in which the marker is located to a band or sub-band. A few gene products and cloned genes have been mapped to a chromosomal band or sub-band by cell hybrid strategy.[9]

It is of considerable importance for site-specific chromosome mapping to apply *in situ* hybridization when possible to map cloned genes and DNA polymorphisms. Of course, a more precise map will stimulate a more specific application of the gene map to human disease.

Following the procedures illustrated in Figure 2, we have mapped by *in situ* hybridization those genes and DNA segments in Table 3.[2,16] This has required using single-copy probes large enough to obtain sufficient hybridization to demonstrate silver grains at the specific chromosomal site. A general size for a probe used in *in situ* studies has not been determined, but success of the procedure does depend on several parameters including the number of gene copies.

Table 4 illustrates human genes and DNA segments mapped by *in situ* hybridization that have been published at this writing. The number of markers mapped in this way has been obtained in a relatively short time and forecasts that a large number of diverse markers will be identified and mapped in this manner.

TABLE 4. Human Markers Assigned by *In Situ* Hybridization[1,2]

Chromosome	Marker	Symbol	Segment
1	Amylase	AMY	1p21
	14.5 kb*	D1S1	1p36
	5S RNA	RN5S	1q42→q43
2	Kappa light chain immunoglobulin	IGKV	2cen→p12
	Proopiomelanocortin	POMC	2p23
3	D3S1*	D3S1	3q12
	Somatostatin	SST	3q28
4	Albumin	ALB	4q11→q13
7	Histones	H1, H2B, H4, H2A, H3	7q2
11	Insulin	INS	11p15→p14
	β-Globin	HBB	11p12.5→p12.8
14	γ^4 Heavy chain immunoglobulin	IGHG4	14q32
16	α-Globin	HBA	16p12→pter
17	Growth hormone	GH	17q22→q24
	Chorionic somatomammotropin	CSH	17q22→q24
13,14,15,21,22	Ribosomal RNA	RNR	p12

*DNA segment

MAPPING GENES TO STUDY DISEASE

For many diseases it is known that the defect is an abnormal enzyme, protein, or other biochemical property. By understanding the genetics of the normal structural gene of the gene product altered in these diseases, the genetics and mapping of the disease can be determined.[42] Under these circumstances, such genes expressed in cultured cells can be mapped using cell hybrids or by *in situ* hybridization if a cloned probe is available. However, only under certain conditions can a disease of unknown etiology, or sometimes even a structural gene mutant, be assigned to a specific chromosome by family studies.

It is important to chromosomally map human disease in order to best understand the disease, understand its genetic organization with respect to adjacent genes and distant controlling genes, predict and prevent the disease, provide genetic counseling for the disease, and study possible genetic therapy.[1,5] For many diseases, especially those of unknown etiology, it is necessary to already have a detailed gene map with markers that can be used in family studies for gene linkage relationships; a disease closely linked to a mapped gene automatically assigns the disorder to a chromosome. The key for genetic dissection and fine resolution mapping of human disease is to know the altered gene for mapping, or to have appropriate, specifically located, polymorphic markers throughout the genome for gene linkage studies.

Table 5 lists the molecular markers we have mapped that are either associated with disease when altered, or that are polymorphic and can be used in family studies. The table also indicates those markers which mapping evidence suggests are closely syntenic to the disease marker.[1,9] Table 5A lists the polymorphic markers and those mapped (often not definitively) diseases that can be used in family studies to determine linkage relationships and more precise mapping. Table 5B lists those non-polymorphic markers that are associated with disease when mutated and are important for disease characterization. These markers can serve as benchmarks to isolate adjacent DNA polymorphisms if specific chromosomal sites are important in the final mapping of diseases. With these markers, a substantial number of significant diseases can be studied genetically.

CONCLUSIONS

Eventually the majority of the human nuclear genome will be sequenced. At that time, genes will be identified by their exact position on a chromosome; perhaps this site will be identified by the starting and ending nucleotide number. In the meantime, the level of high resolution mapping will be limited to localizing a gene to a chromosome band or sub-band. However, this level of resolution will be enormously important for making the human genome extremely useful for the study of human molecular and medical genetics. Site-specific chromo-

TABLE 5. Mapping Genes to Study Disease

Chromosome	Marker	Segment	Associated Disease	Close Syntenic Disease Marker
A. Polymorphic markers to study disease				
1	AMY	p21		Rh erythroblastosis, Elliptocytosis, Galactose epimerase deficiency, Fucosidosis
3	D3S1*	q12		
	SST	q28	Increased in some tumors	Small cell lung carcinoma, GM_1 gangliosidosis, Glutathione peroxidase deficiency
	12-32*	q21→qter		
5	12-65*	pter→q35		GM_2 gangliosidosis and Maroteaux-Lamy syndrome
6	PRL	pter→p23	Prolactin deficiency	HLA, complement component deficiency, Congenital adrenal hyperplasia III
9	12-8*	pter→q12		Galactosemia, Citrullinemia
11	INS	p15→p14	Diabetes (rare form)	Acatalasia, Wilms' tumor-aniridia, Sickle cell anemia, β-Thalassemia
	ADJ-762*	p11→pter		
12	KRAS2		Proto-oncogene	Triosephosphate isomerase deficiency
16	CTRB			Hemoglobinopathies, Hyperuricemia
17	12-2*			Galactokinase deficiency, Pompe's disease
18	12-62*			
21	237D*			Trisomy 21
22	MS3-18*	pter→q13		Ph^1, Methemoglobinemia, Metachromatic leukodystrophy

TABLE 5. continued

Chromosome	Marker	Segment	Associated Disease	Close Syntenic Disease Marker
B. Non-polymorphic Markers				
2	POMC	p23	Ectopic ACTH synd.	Immunoglobulin κ chain disorders?
9	IFA	pter→q12	Immune system dysfunction?	Galactosemia, Citrullinemia
	IFB	pter→q12		
11	PTH	p11→pter	PTH deficiencies	Acatalasia, Wilms' tumor-aniridia, Sickle cell anemia, β-Thalassemia
12	IFG	p12→qter	Immune system dysfunction?	
17	GH	q22→q24	Growth hormone deficiency	Galactokinase deficiency, Pompe's disease
	CSH	q22→q24	CSH deficiency	
20	SRC		Proto-oncogene	Severe combined immunodeficiency due to adenosine deaminase deficiency

*DNA segments

somal mapping of cloned genes, coupled with the more available gene-linkage possibilities provided by the large increase of polymorphic markers, will order genes linearly and determine distances between them. The possibilities for high resolution mapping and genetic dissection of human disease have never been greater. This information should be very important in understanding and managing inherited disease.

ACKNOWLEDGMENTS

The excellent technical assistance and contributions of R. Eddy, L. Haley, M. Byers, and M. Henry are gratefully acknowledged. In the preparation of the manuscript, the invaluable contributions of C. Young are very much appreciated.

This project was supported in part by grants from the National Institutes of Health (GM 20454 and HD 05196), the March of Dimes (1–692), and the American Cancer Society (CD–62).

REFERENCES

1. Shows TB, Sakaguchi AY, Naylor SL. Mapping the human genome, cloned genes, DNA polymorphisms, and inherited disease. *In:* Harris H, Hirschhorn K, eds. Advances in human genetics. Vol. 12. New York & London: Plenum Press, 1982: 341-52.
2. Zabel BU, Naylor SL, Sakaguchi AY, Shows TB. High resolution chromosomal localization of human genes for amylase, proopiomelanocortin, somatostatin, and a DNA fragment (D3S1) by *in situ* hybridization. *Proc Nat Acad Sci USA* (submitted 1983).
3. Shows TB, Sakaguchi AY. Gene transfer and gene mapping in mammalian cells in culture. *In Vitro* 1980; 16:55-76.
4. Potstein D, White RL, Skolnick M, Davis RW. Construction of a genetic linkage map in man using restriction fragment length polymorphisms. *Am J Hum Genet* 1980; 32:314-31.
5. Shows TB. Mapping the human genome and metabolic diseases. *In:* Littlefield JW, De-Grouchy J, eds. Birth defects. Amsterdam: Excerpta Medica, 1978:66-84.
6. Oslo Conference (1981). Human gene mapping 6. *Cytogenet Cell Gene* 1982; 32:1-341.
7. Cook PJL, Hamerton JL. Report of the committee on the genetic constitution of chromosome 1. *Cytogenet Cell Genet* 1982; 32:111-20.
8. Miller OJ, Siniscalco M. Report of the committee on the genetic constitution of the X and Y chromosomes. *Cytogenet Cell Genet* 1982; 32:179-90.
9. Shows TB, McAlpine PJ. The 1981 catalogue of assigned human genetic markers and report of the nomenclature committee. *Cytogenet Cell Genet* 1982; 32:221-45.
10. Tricoli JV, Naylor SL, Bell GI, Rutter WJ, Shows TB. Human amylase genes are located in the p22→p21 region of chromosome 1. (personal communication).
11. Gerhard DS, Kawasaki ES, Bancroft FC, Szabo P. Localization of a unique gene by direct hybridization *in situ*. *Proc Nat Acad Sci USA* 1981; 78:3755-59.
12. Harper ME, Saunders GF. Localization of single copy DNA sequences on G-banded human chromosomes by *in situ* hybridization. *Chromosoma* 1981; 83:431-39.
13. Malcolm S, Barton P, Murphy C, Ferguson-Smith M. Chromosomal localization of a single copy gene by *in situ* hybridization-human β-globin genes on the short arm of chromosome 11. *Ann Hum Genet* 1981; 45:135-41.

14. Owerbach D, Bell GI, Rutter WJ, Shows TB. The insulin gene is located on chromosome 11 in humans. *Nature* 1980; 286:82-4.
15. Owerbach D, Bell GI, Rutter WJ, Brown JA, Shows TB. The insulin gene is located on the short arm of chromosome 11 in humans. *Diabetes* 1981; 30:267-70.
16. Zabel BU, *et al.* High resolution *in situ* hybridization: Localization of a DNA restriction polymorphism, the human proopiomelanocortin gene, and the human insulin gene. *Am J Hum Genet* 1982; 34:153A.
17. Owerbach D,Rutter WJ, Martial JA, Baxter JD, Shows TB. Genes for growth hormone, chorionic somatomammotropin, and growth hormone-like gene on chromosome 17 in humans. *Science* 1980; 209:289-92.
18. Harper ME, Barrera-Saldana HA, Saunders GF. Chromosomal localization of the human placental lactogen-growth hormone gene cluster to 17q22→24. *Am J Hum Genet* 1982; 34:227-34.
19. Owerbach D, Rutter WJ, Cooke NE, Martial JA, Shows TB. The prolactin gene is located on chromosome 6 in humans. *Science* 1981; 212:815-16.
20. Naylor SL, Sakaguchi AY, Shows TB. Human chromosome organization of cloned hormone genes using somatic cell hybrids. Proceedings, 13th International Cancer Congress. Seattle, Washington. 1981; 676.
21. Owerbach D, *et al.* The proopiocortin (adrenocorticotropin/β-lipotropin) gene is located on chromosome 2 in humans. *Somat Cell Genet* 1981; 7:359-69.
22. Nayor SL, Sakaguchi AY, Shen L-P, Bell GI, Rutter WJ, Shows TB. Polymorphic human somatostatin gene is located on chromosome 3. *Proc Nat Acad Sci USA* 1983. (in press)
23. Bishop JM. Oncogenes. *Sci Am* 1982; 246:80.
24. Southern EM. Detection of specific sequences among DNA fragments separated by gel electrophoresis. *J Mol Biol* 1975; 98:503-17.
25. Sakaguchi AY, Naylor SL, Shows TB. A gene homologous to Rous sarcoma virus v-src is on human chromosomes 20. *Progr Nucl Acids Res Mol Biol* (in press 1983).
26. Sakaguchi AY, Naylor SL, Shows TB, Toole JJ, McCoy M, Weinberg RA. Human c-Ki-*ras*2 proto-oncogene on chromosome 12. *Science* 1983; 219:1081-83.
27. Sakaguchi AY. Genetic organization of human proto-oncogenes. *In*: O'Connor TE, Rauscher FJ, "Oncogenes and retroviruses." Progress in Clinical and Biological Research. Vol. 119. New York:Alan R. Liss, Inc., 1983:93-103.
28. Van Dyke DL, Jackson CE, Babu VR. Multiple endocrine neoplasia type (MEN–2): An autosomal dominant syndrome with a possible chromosome 20 deletion. *Am J Hum Genet* 1981; 33:69A.
29. Weinberg RA. Oncogenes of spontaneous and chemically induced tumors. *Adv Cancer Res* 1982; 36:149.
30. Gahrton G, Robert K-H, Friberg K, Julinsson G, Bilberfeld P, Zech J. Cytogenetic mapping of the duplicated segment of chromosome 12 in lymphoproliferative disorders. *Nature* 1982; 297: 513-14.
31. Owerbach D, Rutter WJ, Shows TB, Gray P, Goeddel DV, Lawn RM. Leukocyte and fibroblast interferon genes are located on human chromosome 9. *Proc Natl Acad Sci USA* 1981; 78:3123-27.
32. Shows TB, Sakaguchi AY, Naylor SL, Goeddel DV, Lawn RM. Clustering of leukocyte and firbroblast interferon genes on human chromosome 9. *Science* (in press 1982).
33. Naylor SL, Sakaguchi AY, Shows TB, Law ML, Goeddel DV, Gray PW. Human immune interferon (IFN-γ) is located on chromosome 12. *J Exper Med* (in press 1983).
34. Sakaguchi AY, Naylor SL, Quinto C, Rutter WJ, Shows TB. The chymotrypsinogen B gene (CTRB) is on human chromosome 16. *Cytogenet Cell Genet* 1982; 32:313.
35. Naylor SL, Sakaguchi AY, Shows TB, Grzeschik K-H, Zasloff M. Holmes M. Two nonallelic tRNA ^{met_i} genes are located in the p23→q12 region of human chromosome 6. *Proc Nat Acad Sci USA* 1983 (in press).

36. Bell GI, Selby MJ, Rutter WJ. Sequence of a highly polymorphic DNA segment in the 5′ flanking region of the human insulin gene. *Nature* 1982; 295:31-5.
37. Wyman AR, White R. A highly polymorphic locus in human DNA. *Proc Natl Acad Sci USA* 1980; 77:6754-58.
38. Jeffreys AJ. DNA sequence variants in the ^{G}gamma–, ^{A}gamma–, delta- and beta-globin genes of man. *Cell* 1979; 18:1-10.
39. Skolnick M. Population Genetics of Restriction Fragment Langth Polymorphisms. Banbury Reports. Cold Spring Harbor Laboratory (in press 1982).
40. Naylor SL, Sakaguchi AY, Crzeschik K-H, Shows TB. Regional localization of genetic markers on chromosome 6. (personal communication).
41. Naylor SL, Sakaguchi AY, Barker D, White R, Shows TB. DNA polymorphic loci mapped to human chromosomes 3, 5, 9, 11, 17, 18 and 22. *Proc Nat Acad Sci USA* (submitted 1983).
42. Shows TB. Genetic and structural dissection of human enzymes and enzyme defects using somatic cell hybrids. *In:* Rattazzi MC, Scandalios JG, Whitt GS. eds. Isozymes: Current topics in biological and medical research. Vol. 2. New York: Alan R. Liss, 1977:107-58.
43. Naylor SL, Sakaguchi AY, Gusella JF, Housman D, Shows TB. Mapping of an arbitrary restriction polymorphism (ARP2) to human chromosome 3. *Cytogenet Cell Genet* 1982; 32:302.
44. Tricoli JV, Gusella J, Housman D, Shows TB. Mapping chromosome DNA polymorphisms on human chromosome 21. (personal communication).

ORIGINS OF ANTIBODY DIVERSITY

Henry Huang
Lee Hood

The vertebrate organism has the remarkable ability to respond to immunization with any arbitrarily chosen antigen by synthesizing large quantities of antibodies specific for the antigen. This is achieved by a process called clonal selection. It is based upon the presence of a large diversity of cells, each capable of synthesizing a single kind of antibody. The antigen stimulates only those cells that synthesize antibodies that tightly bind the antigen. The cells are stimulated to proliferate and to synthesize increased amounts of their antibodies that then constitute the specific antibody response. We will review briefly the ontogenetic origins of the enormous diversity of antibodies that an animal can make, and then present a new model for the evolutionary origin of the DNA rearrangements that are central to the generation of antibody diversity.

ANTIBODY MOLECULES AND GENES

The antibody molecule is composed of two polypeptide chains, a larger heavy (H) chain, and a smaller light (L) chain. Each chain is divided into two functionally distinct portions. The amino-terminal portion is called the variable (V) region, since its sequence is extremely variable. The V regions of both H and L chains participate in antigen binding. The carboxyl-terminal portion is called the constant (C) region since only relatively few types of C regions are found in an animal. The C regions serve as structural support for the V regions, they also provide important effector functions. For example, when antigen is bound to the V region, the C regions are triggered to activate processes (*e.g.*, complement activation) that ultimately destroy or eliminate the foreign antigen. It is the C region of heavy chains that determines the class of the antibody, *e.g.*, IgM antibodies have μ heavy chains and IgG have γ heavy chains.

Three multigene families; κ and λ families for the light chain and the H family, encode the antibody polypeptides. Each polypeptide is encoded by several gene segments. L chains are encoded by variable (V_L), joining (J_L) and constant C_L gene segments, H chains are encoded by V_H, diversity (D), J_H and C_H gene

ISBN 0-12-492220-1

TABLE 1. Recognition Site for *V Gene* Formation

	Location relative to gene segment	Type (1 or 2 turn)
V_λ	3′	2
J_λ	5′	1
V_κ	3′	1
J_κ	5′	2
V_H	3′	2
D	5′ & 3′	1
J_H	5′	2

segments. The V_L and J_L, and the V_H, D and J_H gene segments are separate in the germline, and are joined together by DNA rearrangements to form the complete V_L and V_H genes during the differentiation of antibody-producing cells.[1-4] The process of DNA rearrangements that result in the construction of the *V* genes is called *V* gene formation.

There are two conserved sequences, seven and ten base pairs (bp) in length, that are located immediately 3′ to all germline *V* gene segments, 5′ to all germline *J* gene segments, and on both 5′ and 3′ sides of all *D* gene segments examined (Table 1). The location and conservation of these sequences in all three families and in both mouse and man suggest that they are recognized by the enzymes that mediate *V* gene formation.[4-6] The conserved sequences are separated from each other either by 11 bp, *i.e.*, approximately one DNA helical turn (J_λ, V_κ and *D* gene segments), or by 20-30 bp, *i.e.*, approximately two DNA helical turns (V_λ, J_κ, V_H and J_H gene segments). In every case, correct *V* gene formation requires that a gene segment that is flanked by a "1 turn" recognition site to be joined to a gene segment that is flanked by a "2 turn" recognition site (Table 1). This is called the "1 turn to 2 turn" joining rule.[4-7]

There is a second kind of DNA rearrangement involved in the expression of antibody genes. *V* gene formation in the H family places the *V* gene 5′ to the C_μ gene, since all J_H gene segments are 5′ to the C_μ gene.[4,7,8] The cell at this stage synthesizes μ chains, and therefore IgM molecules.[9,10] Subsequently the cell can make μ chains as well as heavy chains of a different class. Part of this can be attributed to alternate RNA processing. Primary transcripts are known to be processed to mRNA's encoding either membrane-bound or secreted IgM molecules.[11] Long transcripts that include the sequence of the V_H gene, the C_μ gene and other C_μ genes 3′ to the C_μ gene may be processed to give mRNA's that encode different classes of H chains.[12] In addition, the cell can rearrange its DNA such that any C_H gene can be placed 3′ to the V_H gene.[7,13,14] This second kind of DNA rearrangement is called class switching.

ONTOGENETIC GENERATION OF ANTIBODY DIVERSITY

Three sources of antibody diversity have been identified.

First, there are multiple germline genes. There are perhaps 100–300 germline V_H and V_κ gene segments, ten or more D gene segments, and four germline J_H and four germline J_κ gene segments in the mouse.[5,7,8,15,16]

Second, the immune system has evolved powerful combinatorial strategies for generating antibody diversity. The gene segments may be joined in a combinatorial fashion to amplify V gene diversity. For example, if any V_κ can join with any J_κ, and any V_H to D, and any D to J_H, then 800 V_κ (200 V_κ x 4 J_κ) and 8000 V_H (200 V_H x 10 D x 4 J_H) genes can be made by combinatorial joining. In addition, since the antigen-binding site of antibody molecules are made up of both V_H and V_L sequences, combinatorial association of H and L chains amplifies the number of different antigen-binding sites that can be made. For example, if every κ chain can associate with every H chain, then 800 V_κ x 8000 V_H = 6.4 x 10^6 antigen-binding sites can be made. Furthermore, each V_H sequence can associate with each of a number of C_H sequences by alternate RNA processing and by class switching. This puts the many possible antigen-binding sites in association with a number of different sets of effector functions.

The third source of antibody diversity involves two types of somatic alteration of V gene sequences. The first is known as junctional diversity[5,6,15] The process of V gene formation is imprecise. The joining of gene segments can occur at a number of positions at the boundaries of the gene segments. The imprecision results in variability in the sequence and the number of codons that are included in the junctions. Although there is a penalty in imprecise joining that results in frame shifts and termination codons, the increased diversity at the V_L–J_L, V_H–D, and D–J_H junctions is functionally significant, since these junctions encode sequences in the antigen-binding site.[17] The second type of somatic alteration of V gene sequences is somatic mutations [18-22] which occur throughout the V_H and V_L genes as well as their flanking sequences. They involve both silent changes and changes that result in amino acid replacement, they are quite extensive (at least 44 substitutions in and around the M167 V_H gene)[21] and they seem to be correlated with class switching, in that V_H and V_L sequences of IgM are predominantly germline, while most of the mutated V sequences are in antibodies of the other classes.[20,23] The significance of somatic mutation as a functional source of antibody diversity is not clear. We have speculated that somatic mutation allows the immune system to fine-tune the antibody response,[20] and that it might also have a predictive function, by facilitating the response to new invasion by foreign organisms that are genetically, and therefore structurally, related to – but not identical to – the original invading organism that stimulated the immune response.[24]

THE EVOLUTIONARY ORIGIN OF ANTIBODY DIVERSITY

The evolutionary origin of some of the mechanisms for generating antibody diversity are unremarkable. Multiple germline genes usually are a source of genetic diversity for genes of higher eukaryotes. Indeed, non-multigenic systems constitute exceptions rather than the rule in these organisms. Similarly, the combinatorial association of dissimilar polypeptide subunits to generate functional diversity is well established in many other systems, the most obvious examples being hemoglobin and lactate dehydrogenase. Recent evidence (*e.g.* *a* Amylase,[25] calcitonin,[26]) indicates that the strategy of alternate RNA processing is also not uncommon.

All the remaining mechanisms for generating antibody diversity, *i.e.*, combinatorial joining, junctional diversity, class switching and somatic mutation, are so far unique to the antibody system (although there is some evidence that the antigenic variation of trypanosomes involves both combinatorial DNA rearrangements as well as "somatic" mutation.[27]) All of these mechanisms depend on the central role of developmentally programmed DNA rearrangements. Therefore, a large part of the problem of the unique aspects of evolutionary origin of antibody diversity centers on the problem of the origin of DNA rearrangements.

As the sequence requirements for class switching are controversial,[28,29] we will not speculate further on its evolutionary origins. The sequence requirements and properties of *V* gene formation are clearer, and we will present two models for its origin.

The existence of DNA rearrangements in a number of organisms is known.[30] The most familiar examples are the large number of different transposable elements found in a number of organisms. This led Sakano *et al*[6] to postulate that the origin of *V* gene formation in the λ family is in the insertion of a transposable element into a primordial *V* gene. The insertion split the primordial *V* gene into two portions: the primordial *V* and *J* gene segments. Expression of the split gene would then require the developmentally programmed excision of the transposable element coupled with the religation of the gene segments. This is an attractive model, since bacterial and eukaryotic transposable elements can indeed insert into, and inactivate or modify the activity of, a gene, and excision can lead to the restoration of gene activity.[31] Furthermore, the gene-regulating activity and transposition of the inserted segment can be regulated in *trans* by linked or unlinked genes (*e.g.*, the Spm system in maize).[32] Thus the transposed segment need not contain its own transposable gene. Even more striking is the movement of some transposable elements that correlate with normal developmental events (*e.g.*, the *crown* and *flow* alleles in the *En* system of maize).[33] Similarly, other developmentally regulated DNA rearrangements such as chromosomal diminution, the amplification of ribosomal genes, and the production of macronuclear DNA in flagellates are well known.[30] Thus the model invokes processes and features of transposable element-mediated DNA rearrangement that are known to be possible. The difficulty of the

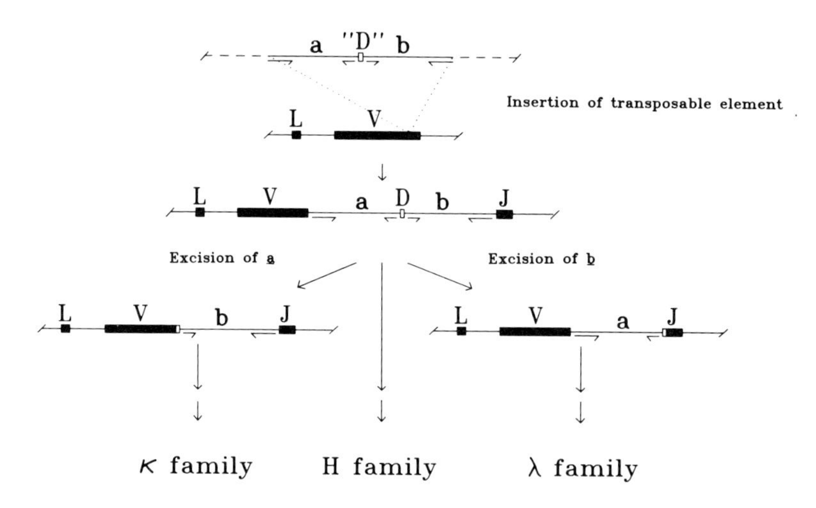

Fig. 1. A model for the origin of *V gene* formation. Insertion of a composite transposable element created the primordial H family. Excision of one or the other insertion sequence in the composite transposable element then gave rise to the primordial κ and λ families. L: leader sequence, V: *V* gene segment, D: *D* gene segment, J: *J* gene segment. Long arrow: 2 turn recognition sequence. Short arrow: 1 turn recognition sequence.

model is that it cannot be generalized to all the antibody families without making multiple *ad hoc* postulates. To account for the different positions of the 1 turn and 2 turn recognition site in the κ and λ families (Table 1), this model requires the insertion of the same or a related transposable element into the κ precursor in the opposite orientation compared to the λ precursor – or alternately, the transposable element in λ must invert itself, to give rise to the κ family. As many as two insertions in both orientations are required to generate the H family configuration. We find this unatractive for the following reasons. First, the *V* and *J* gene segments of the three families are clearly very similar. Thus the model requires that the insertion of the transposable element occurred repeatedly at the same relative location in the primordial gene of each family. Although transposable elements can show specificity for a particular insertion site, witness the integration of *lambda* phage into the *E. coli* genome, we feel that it is an unnecessary complication of the model. Second, the model minimally requires two insertions: one into λ, followed by inversion to give κ, and at least once more to give the H family. The successful insertion of a transposable element into a primordial *V* gene, such that its excision is developmentally regulated and the excision is reasonably precise to restore proper gene activity, involves individual requirements each of which is not too unlikely. However, the successful conjunction of all the requirements must be quite rare. It is rather implausible for it to occur two or more times.

The model can be modified to meet these objectives. Our postulate is that the H family is the first to appear. We think that a composite transposable element, with a structure analogous to that of bacterial transposons, inserted into a primordial V_H *gene* (Figure 1). The transposon has two flanking insertion sequences (IS), labeled *a* and *b* in Figure 1, each bearing a 1 turn and a 2 turn recognition site at its ends. The two IS are each transposable, and are in inverted orientation to each other, properties common in bacterial transposons. They flank a central sequence, analogous to the central antibiotic resistance genes of many bacterial transposons,[31] which is the ancestor of the *D gene* segment. The primordial *V gene* now has the structure V, IS*a*, D, IS*b*, J, exactly as the H family is believed to be organized. The light chain families are derived from the H family by the excision of IS*a* (to give rise to the κ family) or IS*b* (to give rise to the λ family). This model, therefore, postulates a single successful insertion event that is likely to be rare. The subsequent events postulated by the model are direct outcomes of the properties of transposable elements, and can readily occur. We feel that our model is a parsimonious explanation of the origin of the three antibody families, especially since it directly explains the positions of the 1 turn and 2 turn recognition sites in the three families (Table 1). The model invokes frequently observed properties of contemporary transposable elements, and requires only a single primordial event that is likely to be rare. An interesting feature of the model is that it requires the insertion of a transposable element flanked by 2 turn recognition sites on its extreme ends, thus predicting that the insertion of the transposable element need not obey the 1 turn to 2 turn rule, while excision does obey the rule.

ACKNOWLEDGMENTS

This work was supported by a National Institutes of Health grant AI 16913.

REFERENCES

1. Dreyer WJ, Bennett JC. The molecular basis of antibody formation: A paradox. *Proc Natl Acad Sci USA* 1965; 54:864-68.
2. Brack C, Hirama M, Lenhard-Schuller R, Tonegawa S. A complete immunoglobulin gene is created by somatic recombination. *Cell* 1978; 15:1-14.
3. Seidman JG, Leder P. The arrangement and rearrangement of antibody genes. *Nature* 1978; 276:790-95.
4. Early P, Huang H, Davis M, Calame J, Hood H. An immunoglobulin heavy chain variable region gene is generated from three segments of DNA: V_H, *D* and J_H. *Cell* 1980; 19: 981-92.
5. Max EE, Seidman JG, Leder P. Sequence of five potential recombination sites encoded close to an immunoglobulin k constant region gene. *Proc Natl Acad Sci USA* 1979; 76: 3450-54.
6. Sakano H, Huppe K, Heinrich G, Tonegawa S. Sequences at the somatic recombination sites of immunoglobulin light chain genes. *Nature* 1979; 280:288-94.

7. Sakano H, Maki R, Kurosawa Y, Roeder W, Tonegawa S. Two types of somatic recombination are necessary for generation of complete immunoglobulin heavy-chain genes. *Nature* 1980; 286:676-83.
8. Newell N, Richard JE, Tucker PW, Blattner FR. *J genes* for heavy chain immunoglobulins of mouse. *Science* 1980; 209:1128-32.
9. Raff MC. Development and modulation of B lymphoctyes: Studies on newly formed B cells and their putative precursors in the hemopoietic tissues of mice. *Cold Spring Harbor Symp Quant Biol* 1976; 41:159-62.
10. Cooper MD, Kearney JF, Lydyard PM Grossi CE, Lawton AR. Studies of generation of B-cell diversity in mouse, man and chicken. *Cold Spring Harbor Symp Quant Biol* 1976; 41:139-45.
11. Early P, *et al.* Two mRNA's can be produced from a single immunoglobulin μ gene by alternative RNA processing pathways. *Cell* 1980b; 20:313-19.
12. Moor KW, *et al.* Expression of IgD may use both DNA rearrangement and RNA splicing mechanisms. *Proc Natl Acad Sci USA* 1981; 78:1800-04.
13. Honjo T, Kataoka T. Organization of immunoglobulin heavy chain genes and allelic deletion model. *Proc Natl Acad Sci USA* 1978; 75:2140-44.
14. Davis MM, *et al.* An immunoglobulin heavy-chain gene is formed by at least two recombinational events. *Nature* 1980; 283:733-38.
15. Weigert M, Perry R, Kelley D, Hunkapiller T, Schilling J, Hood L. The joining of *V* and *J gene* segments creates antibody diversity. *Nature* 1980; 283:497-99.
16. Kurosawa Y, von Boehmer H, Haas W, Sakano H, Trauneker A, Tonegawa S. Identification of D segments of immunoglobulin heavy chain genes and their rearrangement in T lymphocytes. *Nature* 1981; 290:565-70.
17. Amzel LM, Poljak RJ. Three dimensional structure of immunoglobulins. *Ann Rev Biochem* 1979; 48:961-67.
18. Weigert M, Riblet R. Genetic control of antibody variable regions. *Cold Spring Harbor Symp Quant Biol* 1976; 41:837-46.
19. Bothwell ALM, Paskind M, Reth M, Imanishi-Kari T, Rajewsky K, Baltimore D. Heavy chain variable region contribution to the NP^b family of antibodies: Somatic mutation evident in a γ2a variable region. *Cell* 1981; 24:625-37.
20. Crews S, Griffin J, Huang H, Calame K, Hood L. A single V_H *gene* segment encodes the immune response to phosphorylcholine: Somatic mutation is correlated with the class of the antibody. *Cell* 1981; 25:59-66.
21. Kim S, David M, Sinn E, Patten P, Hood L. Antibody diversity: Somatic hypermutation of rearranged V_H *genes*. *Cell* 1981; 27:573-81.
22. Bothwell ALM, Paskind M, Reth M, Imanishi-Kari T, RajewskyK, Baltimore D.Somatic variations of murine immunoglobulin λ light chains. *Nature* 1982; 298:380-82.
23. Gearhart PJ, Johnson ND, Douglas R, Hood L. IgG antibodies to phosphorylcholine exhibit more diversity than their IgM counterparts. *Nature* 1981; 291-29-34.
24. Huang H, Hood L. The expression of genes. *In*: Yohn DS, Blakeslee JR. eds. Comparative Research on Leukemia and Related Diseases. New York:Elsevier North Holland Inc., 1982; 169-72.
25. Young RA, Hagenbuchle O, Schibler U. A single mouse α-amylase gene specifies two different tissue specific mRNA. *Cell* 1981; 23:451-458.
26. Amara SG, Jones V, Rosenfeld MG, Ong ES, Evans RM. Alternative RNA processing in calcitonin gene expression generates mRNA's encoding different polypeptide products. *Nature* 1982; 298:240-44.
27. Borst P, Cross GAM. Molecular basis for trypanosome antigenic variation. *Cell* 1982; 29:291-303.

28. Hood L, *et al.* Two types of DNA rearrangements in immunoglubin genes. *Cold Spring Harbor Symp Quant Biol* 1981; 45:887-98.
29. Marcu BM, Lang RB, Stanton LW, Harris LJ. A model for the molecular requirements of immunoglobulin heavy chain class switching. *Nature* 1982; 98:87-89.
30. Hood L, Huang HV, Dreyer WJ. The area-code hypothesis: The immune system provides clues to understanding the genetic and molecular basis of all recognition during development. *J Supramolec Struct* 1977; 7:531-59.
31. Calos MP, Miller JH. Transposable elements. *Cell* 1980; 20: 579-595.
32. McClintock B. The control of gene action in maize. *Brookhaven Symp Biol* 1965; 18: 162-84.
33. Peterson PA. Phase variation of regulatory elements in maize. *Genetics* 1966; 54:249-66.

THE GENES ENCODING THE MAJOR HISTOCOMPATIBILITY ANTIGENS

J. G. Seidman
Glen A. Evans
Keiko Ozato
R. Daniel Camerini-Otero
David Margulies

INTRODUCTION

The major histocompatibility antigens are a family of proteins that play a primary role in cell-cell recognition events (for reviews see 1–3).* These molecules, first discovered by P. Gorer in the mouse (called H-2 antigens) and discovered later in humans (called HLA) and other vertebrates, were characterized by several different tests. First, these cell surface proteins are very antigenic so that when cells from one member of a species are injected into another member of the species an alloantiserum is raised that primarily recognizes the histocompatibility antigens. Second, when cells from one individual are introduced into another host of the same species these cells are destroyed (rejected) as the result of small differences in their histocompatibility antigens. The major role played by these antigens in graft-rejection phenomena led to their name – the major histocompatibility antigens.

The major histocompatibility antigens are cell surface glycoproteins consisting of two subunits (Figure 1)[1-3] One subunit, the non-polymorphic, non-glycosylated polypeptide, is encoded on chromosome 17. Reaching this level of understanding of the major histocompatibility antigens has not been easy. These antigens comprise less than 0.1% of the total membrane protein. In fact studies of these proteins have only been feasible with the aid of classical genetic studies.

Studies of the genetics of the region encoding the major histocompatibility antigens have been extensive.[1-3] Snell first showed that the genes encoding the 44,000 dalton subunit of these antigens are all clustered into one small (less than two centimorgans) section of the murine chromosome 17. This small region –

*A similar presentation was made to the Third Annual Meeting of the Transplantation Society and is published in *Transplantation Proceedings.*

ISBN 0-12-492220-1

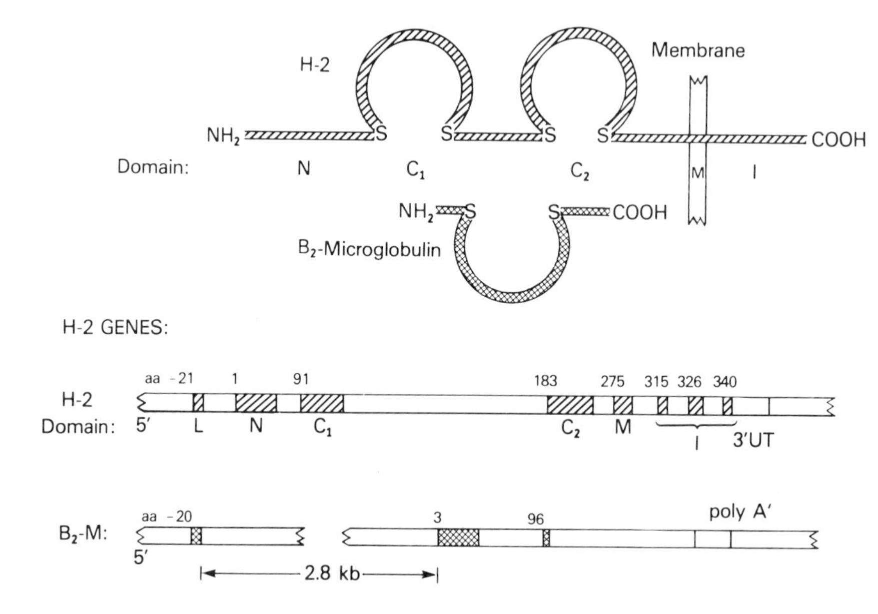

Fig. 1. The structure of H-2 antigens and their genes.

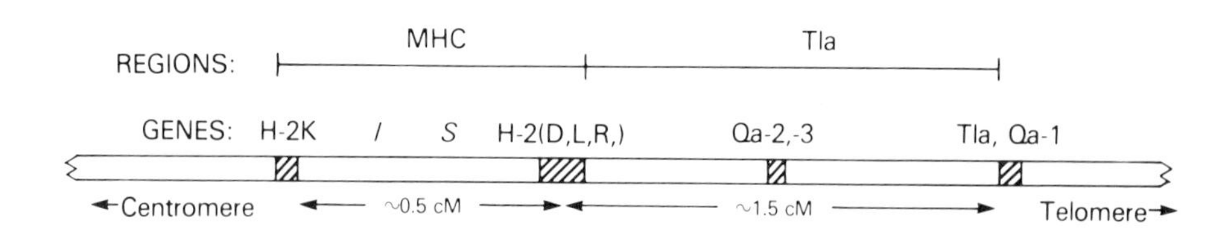

Fig. 2. A classical genetic map of the major histocompatibility complex (MHC) and the thymus leukemia antigen Tla region.

called the major histocompatibility complex and its closely linked *Tla* locus (Figure 2) – is the best studied centimorgan region of any mammalian genome. The intense interest in this region has stemmed from the availability of naturally occurring mutants (due to polymorphism), the relative simplicity of the graft-rejection test, disease associations in humans, and the relationship to immune responsiveness in mice. Despite the extensive study of these antigens many of the original questions remain unanswered. For example, the biological role of these antigens has not been determined. Another puzzle about these antigens has been the very polymorphism that made these studies possible. Most proteins of one inbred mouse line differ by less than 2% from the same proteins of another inbred mouse line. By contrast protein chemistry of the H-2 antigens suggest that the H-2 antigen of one mouse may be as much as 20% different from the H-2 antigen of another mouse. This finding has pointed to the possibility that the evolution of the H-2 antigen genes is different from the evolution of other mouse proteins.

Elucidation of the molecular basis of this polymorphism and the organization of the *H-2* genes awaited the advent of recombinant DNA technology. However, like other studies of these antigens, recombinant DNA technology was originally hindered by the small amounts of antigen made per cell. The first DNA probes to *H-2* gene sequences were not available until about two years ago.[4-6] The availability of these DNA probes has led to the identification of the *H-2* gene family, and the molecular cloning and nucleotide sequence of the *H-2* genes.[7-9] We will review this data and attempt to develop a model to explain the rapid evolution of the H-2 antigens.

BACKGROUND

The extensive polymorphism of the major histocompatibility antigens prevents one member of a species from accepting a tissue graft from another member of the same species. Snell and his collaborators took advantage of this fact to derive congenic strains of mice.[1,2] These strains of mice were constructed so that they are identical to one another throughout their genomes except in the small segment of chromosome 17 that encodes the major histocompatibility antigens. Geneticists realized that the finding that the major transplantation antigen functions appeared to be encoded in a single region did not ensure that these antigens were encoded by only a single gene. Later, the identification of recombinant strains of mice that rejected grafts from both parents demonstrated that at least two closely linked loci, *H-2D* and *H-2K*, encoded major histocompatibility antigens. This genetic approach to the identification of the major histocompatibility antigens was tedious, and led to a minimum estimate of the number of antigens.

Another genetic approach to the H-2 antigen has been to isolate mutant strains of mice that either rejected grafts from their parents or whose skin when transferred to the parents was rejected. These heroic studies performed in several laboratories[10,11] involved the screening of over 50,000 mice by graft-rejection tests. From the twenty mutant strains isolated by this procedure these workers have concluded that the *H-2* genes mutate at the extremely high rate of one mutation per 2,000 haploid genomes.

Biochemical analysis of the H-2 antigens has helped to define the polymorphism of these antigens and the structural alterations in the mutated genes. H-2 antigens are most easily purified by immunoprecipitation with the appropriate alloantiserum. Almost ten years ago several groups demonstrated that these alloantiserum precipitate not one but two polypeptide chains. One of these chains is the polymorphic H-2 chain while the other chain is the non-polymorphic β-2 microglobulin.[1-3,12]

The structure of only one complete mouse H-2 chain and several partial sequences have been determined by protein sequence analysis.[13] Nevertheless, fingerprint analysis and sequence analysis demonstrated that the polymorphism of these chains was extensive. For example, the H-2K^b polypeptide sequence appears to be about 20% non-homologous to the H-2K^d sequence that is available. These sequence analyses appeared to confirm the predictions of the early serologists that the H-2 polypeptides were highly polymorphic.

More recently, several other H-2-like molecules, Qa-1, Qa-2, Qa-3, Qa-4, Qa-5, Qa-6 and Tla antigens have been identified.[14,15] These antigens are somewhat polymorphic, associated with β-2 microglobulin on the cell surface, encoded by genes distal to *H-2D* on chromosome 17 and are 44,000 dalton polypeptides. The homology of these polypeptides to the classical transplantation antigens H-2D or H-2K has not been determined precisely, but the features outlined above have led several groups to propose that all of these antigens are derived from common ancestors.

CLONING H–2 ANTIGEN mRNA's

Over the past two years cDNA clones corresponding to mRNA's encoding both chains of the mouse and human transplantation antigens have been cloned.[4-6,16] The general approach taken to the isolation of these cloned sequences has been to make a cDNA library from a size purified fraction of the polyA+ mRNA of the cell. (A cDNA library contains at least one representative of each mRNA species.) These cDNA libraries were then screened for the desired sequences. In one procedure clones are screened for their ability to select the correct mRNA sequences by hybridization. These selected mRNA's were translated in *in vitro* translation systems. The protein products were immunoprecipitated and identified on SDS-polyacrylamide gels. This approach has been used to clone mouse H-2 cDNA's[6] and β-2 microglobulin,[16] as well as the human HL-A cDNA sequences.[4] A second approach to screening cDNA libraries for H-2 antigen cDNA's has been to use synthetic oligonucleotides either directly as DNA probes, or indirectly as primers to synthesize cDNA probes to identify the desired clones. This approach has been used to clone human β-2 microglobulin[17] and HL-A[5] cDNA sequences.

The major histocompatibility antigens are found on the surface of nearly every cell.[1-3] However, since these antigens comprise less than 0.1% of the total cell protein, the mRNA species are extremely rare. In general the mRNA's correspond to between 0.1–0.01% of the total mRNA. Thus, even after size purification of mRNA between 100 and 1,000 cDNA clones were screened in order to find the desired sequences.

ONE B–2 MICROGLOBULIN GENE AND A FAMILY OF *H–2* LIKE GENES

The number of genes encoding H-2 antigens has been a question ever since geneticists identified the major histocompatibility complex on chromosome 17. Classical genetics can only define the minimum number of genes required to encode a particular phenotype. If a number of very similar genes are encoded at a particular locus these genes can only be distinguished if the genes are polymorphic or if recombinants between these genes have been identified. More recently, gene counting experiments with DNA probes have provided new insights into the organization and size of a number of mammalian gene systems.

The principle of these experiments is to count the number of fragments in a restriction digest of mouse DNA that encode sequences corresponding to a particular DNA probe. An example of the results of a Southern blot experiment where β-2 microglobulin (lane a) and *H-2* (lane b) probes are displayed (Figure 3).[8,18,19]. The striking feature of these results is that while the gene hybridizes to a large number (more than 20) restriction fragments, the β-2 microglobulin probe only hybridizes to a single fragment. The conclusion from experiments of this type is that there is only one β-2 microglobulin gene per mouse genome, but there are more than 20 *H-2*-like genes in the mouse genome.

The presence of only a single β-2 microglobulin gene in the entire mouse genome suggests that the products of this one gene must service the products of all the H-2-like genes. The existence of human[18] and mutant mouse[19] cell lines that do not express β-2 microglobulin and thus cannot express the major transplantation antigens confirm the hypothesis that β-2 microglobulin expression is required for the expression of these antigens. Perhaps the fact that this one gene product must service a variety of H-2 gene products has provided an important selective force on the products of this gene preventing the development of a polymorphic set of β-2 microglobulins.

The finding of a single β-2 microglobulin gene contrasts sharply with the finding of a large number of *H-2* genes. Presumably, the fact that there are many *H-2*-like genes has played an important role in generating the polymorphism of the H-2 antigens. For the scientist the finding of a large family of *H-2* genes has presented a technical problem. One cannot be certain from Southern blot analysis alone which restriction fragment encodes which gene.

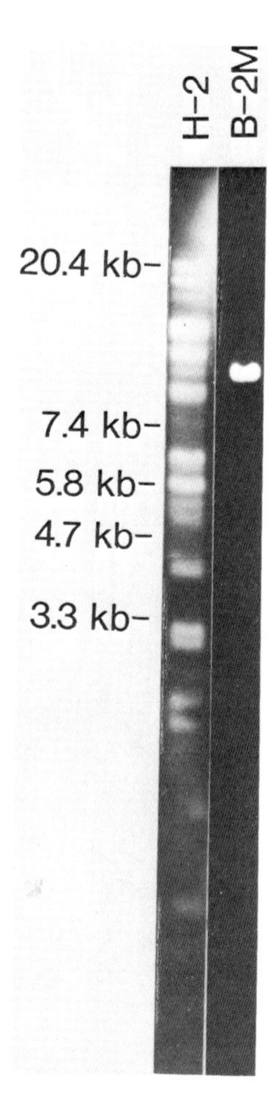

Fig. 3. Restriction fragments encoding β-2 microglobulin and *H-2* genes identified by Southern blot hybridization. The methods for obtaining these Southern blots are described in reference 8.

THE STRUCTURE OF THE *H–2* AND B–2 MICROGLOBULIN GENES

Procedures for determining the nucleotide sequence of a eukaryotic gene are now well defined. Genomic clones are obtained by screening libraries of bacteriophage encoding segments of mouse DNA with a cDNA probe. Using this procedure β-2 microglobulin[19] and H-2[9] bacteriophage clones were isolated. The nucleotide sequence of the genes encoded by these recombinant phage was then determined and the structure of the genes was defined by comparison of the protein and cDNA sequences to the gene sequence.

Analysis of the β-2 microglobulin gene was relatively straightforward. Since only a single restriction fragment of mouse DNA hybridizes to the cDNA probe, one could be certain that the gene cloned on a particular bacteriophage must encode the β-2 microglobulin gene. The location of the intervening sequences within this gene are indicated in Figure 1, lower panel.[19]

The H-2 gene system has presented a more difficult problem. Certainly, particular H-2-like genes could be isolated from bacteriophage libraries,[7,9,20] and the structures of these genes could be determined. The structures of at least three such H-2-like genes have been determined. The overall structure of these genes seems to be similar (Figure 1, lower). The hydrophobic leader segment, each of the protein domains and the hydrophobic membrane piece are all encoded by separate coding blocks.

Despite the fact that *H-2*-like genes can be cloned from phage genomic libraries, a difficult problem remains. The hybridizing DNA sequences must be correlated with a cell-surface antigen. To date two approaches have been used to identify a particular cloned gene. One approach involves determining the nucleotide sequence to the known amino acid sequence. Unfortunately, amino acid sequence data for most of the H-2-like antigens is not yet available. Thus, these sequence comparisons frequently do not allow positive identification of the cloned DNA. A second approach to identifying the cloned DNA is to transform these genes back into mouse L-cells and look for the expression of a new cell surface antigen with monoclonal antibodies.

Both nucleotide sequence analysis and expression of the cloned genes in mouse fibroblasts have been used to characterize the mouse $H\text{-}2L^d$ genes. Several groups[9,21] have isolated cloned DNA's that, when introduced into mouse L-cells, will express a cell surface antigen that is recognized by monoclonal antibodies directed against H-2L^d. Eventually, extensive gene mapping studies using cosmid vectors as begun by several groups[20,22] will locate all of these *H-2*-like genes. The relationship of the different *H-2* genes will be better understood from these studies.

GENETIC EXCHANGE BETWEEN HOMOLOGOUS GENES

The *H-2* gene family comprises more than 15 different closely related genes. Most, if not all, of these homologous genes are encoded on chromosome 17 and are linked to one another. The fact that this is a gene family and the close linkage of the family members to one another must affect the evolution of these genes. Members of a large gene family can probably exchange information and thus generate new gene sequences at a high rate. Evidence for this exchange of genetic information can be derived from an analysis of the mutations that have been identified in some of the *H-2b* mutant strains.

The strains of mice bearing mutationally altered $H\text{-}2K^b$ genes have been studied extensively by Nathenson and his collaborators.[22] A summary of these conclusions are indicated in Table 1. These workers have examined the sequence of the $H\text{-}2K^b$ protein in ten different independently derived mutant strains. Radiolabelled $H\text{-}2K^b$ protein was isolated from each strain and a tryptic fingerprint of each protein was prepared. Comparison of these fingerprints to the fingerprints of the parental $H\text{-}2^k$ molecule revealed one or two tryptic peptides that had been altered in the mutant strains. The altered tryptic peptides were subjected to peptide sequence analysis. The sequence alteration identified in the mutant $H\text{-}2K^b$ proteins are indicated in Figure 4 (top). Surprisingly, most of the alterations could not be accounted for by a single nucleotide substitution. Some of the mutations even involved two amino acid substitutions. For example, the $H\text{-}2^{bm1}$ mutation involved a double substitution of an arg-leu (amino acids 155 and 156) and a tyr-tyr dipeptide (Table 1). This particular mutation requires a minimum of three alterations of the DNA sequence (Figure 4). Furthermore, some mutations were isolated independently. For example, three mutant strains were isolated independently that had mutations like those in the $H\text{-}1^{bm6}$ strain. These patterns of mutational alteration appeared to be consistent with the notion that these new sequences were derived from other *H-2*-like genes.

Structural studies of the $H\text{-}2L^d$ gene demonstrated that these sequences found in the mutant $H\text{-}2K^b$ molecules are derived from *H-2L*-like genes. That is, each of the mutationally altered $H\text{-}2K^b$ gene sequences obtained a sequence that is present in an $H\text{-}2L^d$-like gene. For example, bml has a new sequence at residues 155 and 156. Two tyrosine residues replace the arg-leu sequence of the parent. This tyr-tyr sequence is also found in $H\text{-}2L^d$. In fact, all of the mutations of the $H\text{-}2K^b$ molecule could have been derived from the $H\text{-}2L^d$ gene except for bm10 (Table 1).

There appear to be at least two mechanisms by which these mutationally altered genes could have derived their new sequences (see Figure 4). One model might be that two closely related genes could undergo reciprocal recombination and thus exchange genetic information. Another model proposes that the exchange of genetic information can occur *via* a gene conversion event. Gene conversion events in yeast provide a precedent for this type of event, even though the molecular basis for these events is poorly understood. Regardless of the mechanism involved, we suppose that such events occur as a consequence of the fact that the *H-2* family is large and that the genes are closely related.

An interesting contradiction is seen if one attempts to explain the polymorphism of the H-2 gene in terms of recombination or gene conversion. Other multigene families exist that are not polymorphic. For example, the histone genes are encoded by large families of tandemly duplicated genes. These histone genes are very homogeneous and are not polymorphic, presumably because mutations in the histone gene family are eliminated by unequal crossing over events within the locus. The difference between the polymorphic *H-2* gene family and the non-

TABLE 1. Mutations in the $H\text{-}2K^b$ Gene and their Relationship to Sequences Found in $H\text{-}2L^d$*

Mutant	Position	Substitution		$H\text{-}2L^d$
bm 3	77	Asp	x	Glu
	89	Lys	x	Ala
bm 11	77	Asp	x	Glu
bm 1	155	Arg	Tyr	Tyr
	156	Leu	Tyr	Tyr
bm 5	116	Tyr	Phe	Phe
bm 16	116	Tyr	Phe	Phe
bm 6	116	Tyr	Phe	Phe
	121	Cys	Arg	Arg
bm 7	116	Tyr	Phe	Phe
	121	Cys	Arg	Arg
bm 9	116	Tyr	Phe	Phe
	121	Cys	Arg	Arg
bm 8	23	Met	x	Ile
bm 10	165	Val	Met	Val

*Taken from references 22 and 9.

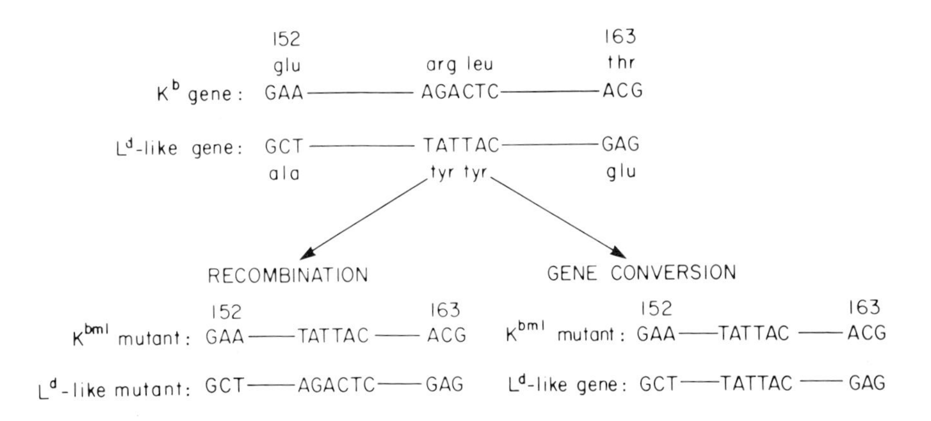

Fig. 4. Two possible mechanisms for genetic exchange between closely related *H-2* genes that could generate the $H\text{-}2^{bm1}$ mutation.

polymorphic histone gene family seems to reside in the spacer sequences between the genes. The histone genes are closely linked (less than 5 kilobases) and the linker regions are highly conserved. In contrast, the regions between one *H-2* gene and the next are much longer by 10 to 30 kilobases of DNA that can be unique in the mouse genome, and even encode other unique genes (for example the immune response genes, Figure 2). Thus, the *H-2* gene family is different from families of tandemly encoded genes.

In many ways the *H-2* gene family closely resembles the immunoglobulin variable region genes. The variable region genes are encoded in sub-families, they are closely linked to one another, but the spacer regions between the genes are not highly conserved. Both of these gene systems seem to be designed to generate further diversity during the course of evolution. Future studies will define these mechanisms for generating and maintaining polymorphism.

SUMMARY

The major histocompatibility antigens play a critical role in the recognition of self and non-self. These antigens, originally thought to be encoded by two genes, now appear to be members of a large family of genes. One source of the polymorphism of these antigens is gene conversion or reciprocal recombination events between members of this multigene family. These events lead to rapid diversification of existing histocompatibility antigen genes.

REFERENCES

1. Klein J. Biology of the Mouse Histocompatibility Complex. Berlin:Springer, 1975.
2. Snell GD, Dausset J, Nathenson, SG. Histocompatibility. New York:Academic Press, 1976.
3. Klein J. The major histocompatibility complex of the mouse. *Science* 1979; 203:516-21.
4. Pleough HI, Orr HT, Strominger JL. Molecular cloning of a human histocompatibility antigen cDNA fragment. *Proc Natl Acad Sci USA* 1980; 77:6081.
5. Sood AK, Pereira D, Weissman SM. Isolation and partial nucleotide sequence of a cDNA clone for human histocompatibility antigen HLA-B by use of an oligodeoxynucleotide primer. *Proc Natl Acad Sci USA* 1981; 78:616.
6. Cami B, Bregegere F, Abastado JP, Kourilsky P. Multiple sequences related to classical histocompatibility antigens in the mouse genome. *Nature* 1981; 291:673.
7. Steinmetz M, Moore KW, Frelinger JG, Sher BT, Shen FW, Boyse EA, Hood LA. A pseudogene homologous to mouse transplantation antigens: Transplantation antigens are encoded by eight exons that correlate with protein domains. *Cell* 1981; 25:683.
8. Margulies DH, Evans GA, Flaherty L, Seidman JG. *H-2*-like genes in the Tla region of mouse chromosome 17. *Nature* 1982; 295:168.
9. Evans GA, Margulies DH, Camerini-Otero RE, Ozato K, Seidman JG. Structure and expression of a mouse major histocompatibility antigen, H-2L^d. *Proc Natl Acad Sci USA* 1982; 79:1994.
10. Bailey DW, Kohn HJ. Inherited histocompatibility changes in progeny of irradiated and unirradiated inbred mice. *Genet Res Camb* 1965; 6:330-40.
11. Melvold RW, Kohn HI, Dunn GR. History and geneology of the *H-2*K^b mutants from the C5+BL/BKl colony. *Immunogenetics* 1982; 15:177-85.

12. Michaelson J, Rothenberg E, Boyse EA. Genetic polymorphism of murine β-2 microglobulin detected biochemically. *Immunogenetics* 1980; 11:93-5.
13. Nathenson SG, Uehara H, Ewenstein BM, Kindt TJ, Coligan JE. Primary structural analysis of the transplantation antigens of murine H-2 major histocompatibility complex. *Ann Rev Biochem* 1981; 50:1025-52.
14. Flaherty L. The TLA region antigens. *In:* Dorf, ME. The Role of the Major Histocompatibility Complex in Immunobiology. New York:Garland STPM, 1980:33-58.
15. Flaherty L. Genes of the TLa region: The new Qa system of antigens. *In*: Morse HC, ed. Origins of Inbred Mice. New York:Academic Press, 1980:409-22.
16. Parnes JR, *et al.* Mouse β-2 microglobulin cDNA clones: A screening procedure for cDNA clones corresponding to rare mRNA's. *Proc Natl Acad Sci USA* 1981; 78:2253-57.
17. Reyes A, Schold M, and Wallace RB. The complete amino acid sequence of the murine transplantation antigen H-2D^{d} as deduced by molecular cloning. *Immunogenetics* 1982; 16:1-9.
18. Michaelson J. Genetic polymorphism of β-2 microglobulin (B2m) maps to the the H-2 region of chromosome 2. *Immunogenetics* 1981; 13:167-71.
19. Parnes JR, Seidman JG. Structure of wild-type and mutant mouse β-2 microglobulin genes. *Cell* 1982; 29:661-69.
20. Steinmetz M, Winoto A, Hinard K, Hood L. Clusters of genes encoding mouse transplantation antigens. *Cell* 1982; 28:489.
21. Goodenow RS. Identification of a BALB/c *H-21*^{d} gene by DNA-mediated gene transfer. *Science* 1982; 215:677.
22. Nairn R, Yamaga K, Nathenson SG. Biochemistry of the gene products from murine MHC mutants. *Ann Rev Genet* 1980; 14:241-77.

THE ORGANIZATION OF HUMAN MAJOR HISTOCOMPATIBILITY CLASS I GENES AS DETERMINED WITH DELETION MUTANTS

Harry T. Orr
Robert DeMars

The major histocompatibility complex (MHC) (the HLA system in man and the H-2 system in mouse) encodes three classes of proteins.[1] Class I proteins are the major transplantation antigens HLA-A, HLA-B and HLA-C. Class II proteins are the immune response related antigens HLA-D/DR, while class III proteins are some components of complement. Class I antigens are glycosylated membrane proteins associated with β_2 microglobulin (β_2m) on the surface of most somatic cells. A feature of the class I gene family is its high degree of polymorphism.[2] This polymorphism is manifested by the large number of alleles at each locus and by the relatively even distribution of alleles within the population. Primary structural analyses at the protein[3-5] and DNA levels[6] show a clear evolutionary relationship between class I antigens, β_2m and immunoglobulin (Ig) constant domains. As in Ig genes, the suggested protein domains of class I molecules[7] are encoded in separate exons.[8-10]

Several laboratories have cloned MHC class I genes of man[9,11,12] and mouse.[6,8.10,13-15] Southern blot analyses using genomic and cDNA probes have detected multiple hybridizing fragments within the murine[6,16,17] and human[18,19] genomes, indicating that class I genes constitute a multigene family. The multiplicity of class I genes and the scarcity of protein sequence information for class I antigens has made the straight forward identification of most clones by sequencing difficult. Serological analysis of mouse L cells transformed with class I genes will help to overcome the problem of clonal identification.[10,20] However, in contrast to the murine H-2 system, genetic analysis of the HLA complex is encumbered by the lack of congenic strains and by the small number of recombinants within the MHC of well-studied families. Therefore, it is difficult to genetically map DNA clones of HLA genes that do not have known phenotypes and alleles. One means of overcoming the problems encountered in mapping the DNA of the human MHC involves probing HLA loss mutants with HLA-specific DNA probes.[19]

MHC non-expressing variants have been isolated from somatic cells of mouse (*e.g.* 21, 22) and man (*e.g.* 23-25). A large diverse collection of MHC loss mutants

ISBN 0-12-492220-1

TABLE 1. The HLA Phenotype of LCL 721 and Mutants

	GLO2	*DR3*	*B8*	*A1*	*GLO1*	*DR1*	*B5*	*A2*
721	+	+	+	+	+	+	+	+
.4, .5, .6, .9, .10, .16 .18, .20, .25, .33, .35	+	+	–	+	+	+	+	+
.64, .74, .80, .87, .108, .109	+	+	+	+	+	+	+	–
.19	+	–	–	–	+	+	+	+
.45.1	–	–	–	–	+	+	+	+
.127	+	+	+	+	–	–	–	–

has been isolated from the human lymphoblastoid cell line LCL 721.[26] LCL 721 is heterozygous at each known HLA locus and at the linked GLO locus on the short arm of chromosome No. 6: *GLO1; HLA-DR1; B5; A2/GLO2; HLA-DR3; B8; A1*. Initially, the selection of HLA-loss mutants after γ-radiation was with an anti-HLA-B8 serum plus complement.[26] All mutants isolated no longer expressed B8 and 61% of them no longer expressed one or more genes *cis*-linked to *B8*. Such multiple loss mutants included some that had lost B8 and A1, B8, A1 and DR3, or B8, A1, DR3 and GLO2. In no case was loss of genes *trans* to *HLA-B8* detected. The number and variety of HLA loss mutants were then increased to include those with mutations affecting the other haplotype alone[27] or that were A-null, B-null, A-null; B-null and HLA-DR null mutants (Table 1) (DeMars *et al*, submitted for publication).

Evidence that the loss of HLA expression is associated with absence of specific MHC DNA has been presented for several γ-ray-induced loss mutants. Mutant 721.19 (an A1, B8 and DR3 loss mutant) has a microscopically detectable interstitial deletion in the short arm of one chromosome, No. 6,[26] where HLA genes have been mapped.[28] DNA hybridization studies showed that mutants 721.19 and 721.52, an A2, B5 and DR1 loss mutant, have lost HLA class I DNA hybridizing fragments.[19] In addition, the appearance of a new hybridizing band in mutant 721.4, a B8 loss mutant, is consistent with a deletion of MHC DNA.[19]

The present report extends the Southern blot analysis of HLA loss mutants derived from LCL 721. Through the use of 5′ and 3′ specific HLA class I cDNA probes and additional HLA loss mutants it is now possible to map about 30% of the *PvuII* generated class I hybridizaing fragments within the HLA complex. The A1, B8 and A2 genes from LCL 721 can be assigned to specific fragments of DNA. In addition, evidence is presented that class I-like genes are located on the telomeric side of *A2*.[39]

EXPERIMENTAL PROCEDURES

Materials

Restriction enzymes were purchased from Bethesda Research Laboratories.

Probes were made by nick-translations[29] with translation kits supplied by New England Nuclear and Amersham. Proteinase K was purchased from Boehringer Mannheim and RNase A was purchased from Sigma.

Mutant Isolation

The lymphoblastoid cell line LCL 721 was established by infecting peripheral blood lymphocytes with Epstein-Barr virus. HLA loss mutants of LCL 721 were selected after γ-radiation (300R) using alloantisera and complement. Procedures used for selection, cloning, karyotyping and phenotypic analysis of HLA loss mutants were as described.[26]

Peripheral Blood Lymphocyte Isolation

Fifty ml of blood were drawn from normal volunteers selected because of their HLA type. Six ml of citrate were used as an anti-coagulent. PBL's were isolated using the Ficoll-Hypaque method[30] and DNA was prepared on the same day.

DNA Isolation

Genomic DNA from LCL 721, the HLA loss mutants and PBL's was purified using a modification of the method of Blin and Stafford.[31] Briefly, cells were washed three times in TEN buffer (10 mM TRIS-HCL pH 7.8, 150 mM NaC1plus 25mM EDTA) and resuspended in the same buffer to 10^8 cells/ml. An equal volume of 3% Sarkosyl, 50 mM EDTA, lmg/ml Proteinase K was added and the cells incubated at 55°C overnight. Then RNase A (heat treated for 30 minutes at 80° C) was added to 50 μg/ml and incubated for one hour at 55°C. The DNA was then extracted twice with phenol for two hours at 4°C, extracted twice with chloroform: isoamyl alcohol (24:1) and dialyzed versus TEN until the OD_{260} of the TEN reached 0. DNA was then ethanol-precipitated and redissolved (1.0 mg/ml) in $T_{10}E_1$ buffer (10 mM TRIS-HCL pH 7.5, 1 mM EDTA).

Analysis of Genomic DNA

Fifteen μg of DNA was digested with either *PvuII* or *HindIII* using buffer conditions suggested by the supplier. Thirty units of each enzyme were added at 0 hour, then 10 units at 3 hours of incubation. Total incubation time was for 5 hours at 37°C. Five microliters of TAE sample buffer (40 mM Tris-acetate pH 7.9, 5 mM sodium acetate, 1 mM EDTA) containing 10% ficoll, 0.1% Bromophenol blue, and 0.1% xylene cyanole FF were added and the DNA separated on a 0.7% agarose gel containing 1 μg/ml ethidium bromide with TAE as the running buffer at 1 v/cm. Prior to blotting, gels were treated in 0.1 M HC1 for 15 minutes at 25°C, 0.5 M NaOH plus 1.5 M NaC1 for 15 minutes at 25°C and 1M Tris-HC1 pH 7.5 plus 3 M NaC1 for 45 minutes at 25°C. Blotting was according to Southern.[32] The filters were then baked at 80°C in a vacuum oven for 6 hours. Prehybridization and hybridization was as described by Wahl *et al.*[33] using 1–2 x 10^6 cpm per/ml hybridization solution of cDNA labelled to a specific activity of 2–5 x 10^8 cpm per μg by nick-translation.[29] After 16 hours of hybridization at 45°C, filters were washed in 4 changes of 250 ml each 0.2x SSC plus 0.5% SDS at 25°C and 4 changes of 250 ml of each of 0.2x SSC at 45°C. Hybridizing bands were visualized by autoradiography using Kodak XAR-5 film and DuPont lighting-plus intensifying screens. *HindIII* digested *lambda* DNA was used as molecular weight markers.

cDNA Probes

The two HLA class I cDNA clones pHLA-1,[11] 500 bp, and the nearly full-length 1,400 bp cDNA clone isolated by Sood *et al*[12] plus a 156 bp *SstII* subclone of the latter, corresponding codon 47 to 111, were used as probes. Prior to labelling, inserts were removed and purified by acrylamide gel (5%) electrophoresis. pHLA-1 does not extend to the poly A tail or poly A addition site. The Sood *et al*[12] cDNA clone extends from the codon for amino acid 39 to 15bp beyond the poly A addition site of the HLA-B7 mRNA molecule (S. Weissman, personal communication).

RESULTS AND DISCUSSION

Phenotypes of LCL 721 and its loss mutants

Table 1 is a compilation of the mutants we used and compares their phenotypes to that of the parental LCL 721. The manner in which the mutants were isolated will be described in DeMars *et al* (submitted for publication).

HLA class I cDNA hybridizing fragments in LCL 721

Restriction enzymes *HindIII* and *PvuII* were used to digest genomic DNA of LCL 721. Southern blots of these digests were then probed with three kinds of HLA class I cDNA clones: (1) the nearly full length B7 cDNA clone isolated by Sood *et al*,[12] (2) a *SstII* subclone of this clone containing the 156bp of coding sequence (codon 47 to codon 111) and (3) pHLA-1, a 500bp cDNA clone from the 3′ end of an HLA class I mRNA.[11] The 3′ clone contains coding sequence for only the carboxy terminal 46 amino acids plus a portion of the 3′ untranslated region. This 3′ clone does not contain the poly A tail or poly A addition site.

The hybridization patterns obtained from the *HindIII* digest of LCL 721 DNA using these HLA cDNA probes (Figure 1) show that this enzyme cuts relatively infrequently within HLA class I genes. Nine out of 15 hybridizing bands detected with the nearly full length probe (Figure 1b) hybridized with both the 3′ (Figure 1a) and 5′ (Figure 1c) specific probes indicating that these fragments are likely to contain complete HLA class I genes. In contrast, none of the fragments that were generated by *PvuII* digestion of LCL 721 DNA and that hybridized with

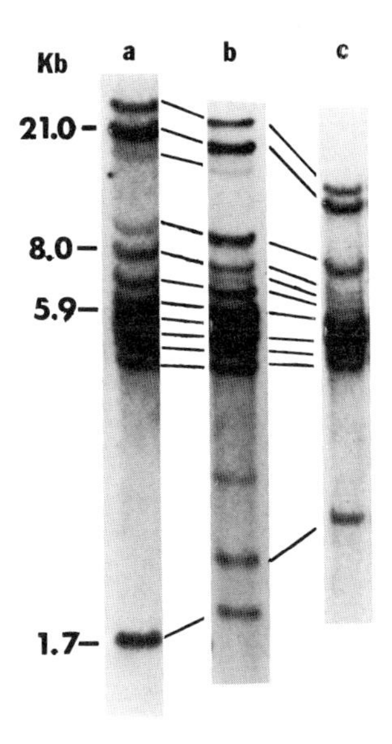

Fig. 1. Southern blot of *HindIII* digested LCL 721 DNA using 5′ and 3′ specific HLA class I probes. Lane a, DNA probed with a 5′ 600 bp *PvuII* fragment of the Sood *et al*[12] cDNA clone containing only coding sequences. Lane b, DNA probed with the Sood *et al*[12] nearly full length class I cDNA clone. Lane c, DNA probed with the 3′ specific probe pHLA–1.[11]

the nearly full length cDNA clone strongly hybridized with both the 3′ and 5′ probes (Figure 2). Three *PvuII* fragments (15 kb, 7.6 kb and 6.2 kb) hybridized the 5′ probe strongly and the 3′ probe very weakly. The two intense 3′ hybridizing fragments (1.05 kb and 0.56 kb) did not hybridize the 5′ probe. Moreover, the simplicity of the 3′ hybridizing pattern compared to that obtained with the 5′ specific probe indicates that the 3′ *PvuII* sites are highly conserved between HLA class I genes while the 5′ *PvuII* sites appear to be more polymorphic.

Distribution of polymorphic fragments between haplotypes of LCL 721

HLA mutants that no longer express one or the other MHC haplotype were used to detect polymorphism in the placement of restriction endonuclease cut sites among class I DNA sequences. Such polymorphism should result in digest fragments that are present in DNA of parental cells and of one kind of haplotype loss mutant but not the other.

Probing *HindIII* digests of LCL 721 and the haplotype loss mutants 721.45.1 and 721.127 with the nearly full length cDNA clone shows that each haplotype

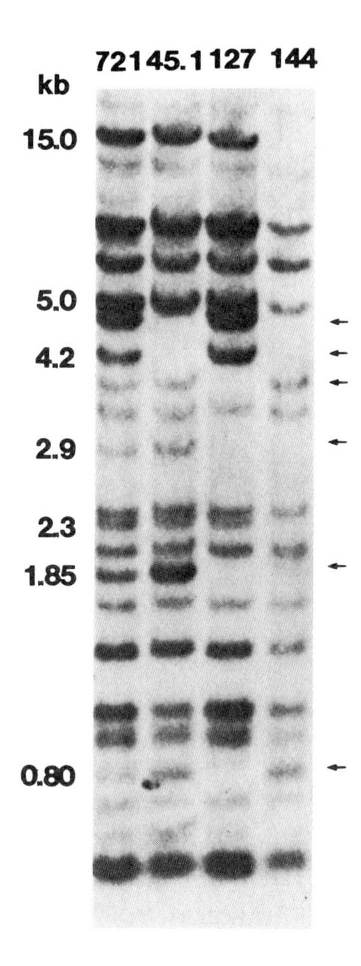

Fig. 2. Southern blot of *PvuII* digested LCL 721 DNA using 5′ and 3′ specific HLA class I probes. Lane a, 156 bp 5′ (codon 47-111) specific probe isolated from the probe used in Lane b. Lane b, the nearly full length class I cDNA probe.[12] Lane c, the 3′ specific probe, pHLA–1.[11] Arrow heads indicate faintly hybridizing band detectable on the original autoradiograph.

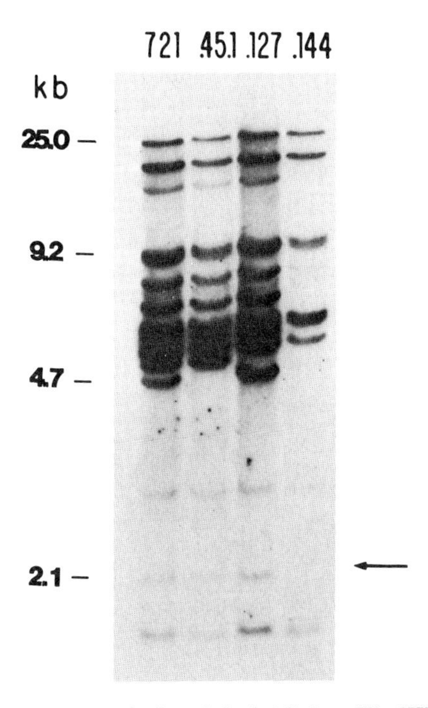

Fig. 3. Haplotype distribution of class I hybridizing *HindIII* generated bands in DNA from: 721, parental LCL; .45.1, *DR3; B8; A1* haplotype loss mutant; .127, *DR1; B5; A2* haplotype loss mutant. The blot was probed with the Sood *et al*[12] nearly full length cDNA clone. Arrow heads indicate bands lost with the deletion of the respective haplotypes.

has only one class I polymorphic *HindIII* fragment (Figure 3) confirming the earlier results obtained with the 3′ specific probe.[19] 721.45.1, the *GLO2-A1* haplotype loss mutant, has lost the 4.7 kb hybridizing fragment while mutant 721.127, the *GLO1-A2* haplotype loss mutant, no longer has the 5.1 kb fragment (Figure 3). The 4.7 kb fragment from the *DR*3-A*1* haplotype and the 5.1 kb fragment of the *DR1-A2* haplotype hybridize to both 5′ and 3′ specific probes. Moreover, Malissen *et al*[9] have recently isolated a complete HLA class I genomic clone located on a 5.6 kb *HindIII* fragment. Thus, it is likely that the 4.7 kb and 5.1 kb bands lost in mutants 721.45.1 and 721.127, respectively, contain complete HLA class I genes.

Restriction enzyme *PvuII* greatly increased the number of polymorphic fragments detectable with mutants 721.45.1 and 721.127 (Figure 4). A total of six polymorphic class I *PvuII* fragments were detectable; two in the *GLO2-A1* haplotype (4.8 kb and 4.2 kb) and four in the *GLO1-A2* haplotype (3.8 kb, 2.9 kb,

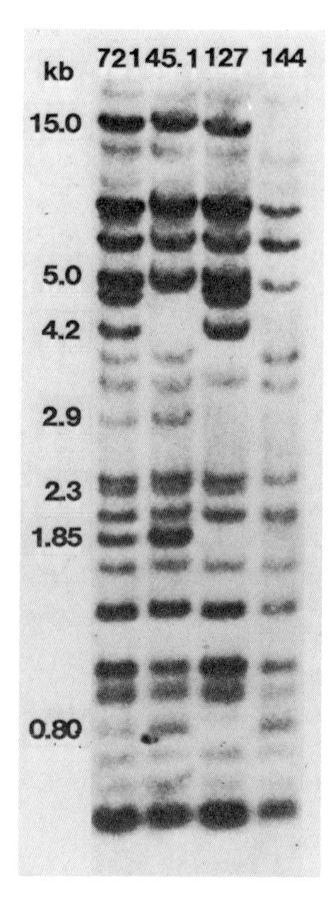

Fig. 4. Haplotype distribution of class I hybridizing *PvuII* generated bands. See legend to Figure 3 for details.

1.85 kb and 0.8 kb). None of these polymorphic *PvuII* fragments hybridized the 3′ specific probe. One polymorphic band at 1.85 kb hybridized only the nearly full length HLA cDNA clone and must represent an internal segment of an HLA gene. The remaining five polymorphic *PvuII* generated bands hybridized only the 5′ specific probe, indicating further that most *PvuII* cut site polymorphisms are located in the 5′ end of HLA class I genes. The relative map positions of these polymorphic *PvuII* bands within the HLA complex indicate that they are generated from at least two class I genes (see below).

Correlation of DNA deletions with the loss of HLA-B8 expression

DNA's from ten B8 loss mutants (Table 1) were digested with *PvuII*, Southern blotted and probed with the nearly full length HLA cDNA clone (Figure 5). DNA from the *B8* haplotype loss mutant 721.19 was also analyzed. In five of the

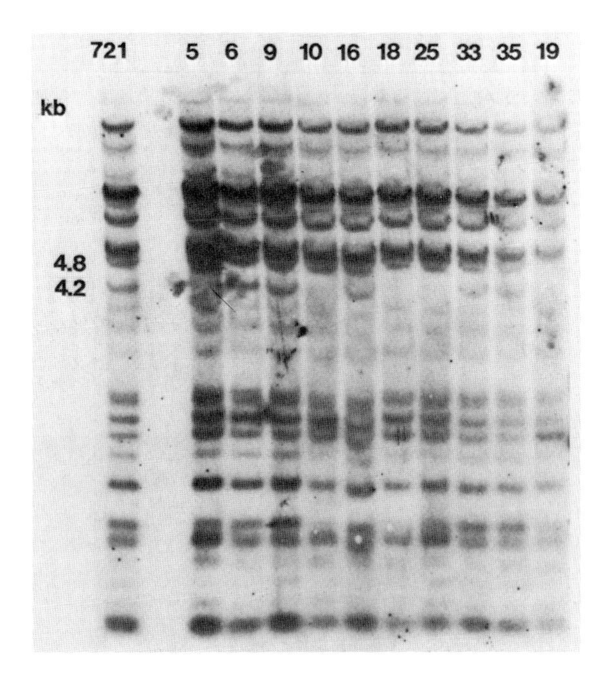

Fig. 5. Disappearance of a 4.2 kb *PvuII* generated band in *HLA–B8* single loss mutants. Mutant DNA was digested with *PvuII* and probed with the Sood *et al*[12] cDNA clone. All lanes contain digested DNA from *B8* single loss mutants, except lane 19 which is from the *DR3; B8; A1* loss mutant.

ten *B8* loss mutants the only change in the Southern blot pattern was the absence of a 4.2 kb band that was present in the DNA of parental LCL 721 cells. The remaining five *B8* loss mutants did not show any loss or gain of hybridizing bands compared to LCL 721.

There are several possible explanations for why some *B8* loss mutants do not seem to have an alteration in their hybridization patterns. A deletion could be too small to detect. However, for most bands deletions as small as 100bp should be detectable. Another possibility is that deletions may have occurred in a segment of DNA that is important for gene expression but not homologous to the cDNA probe used. Mutation resulting in the loss of antigen expression may not have been due to a deletion of DNA. Although principally a deletion mutagen,[34] γ-irradiation may induce point mutations, (*e.g.* 35). Moreover, certain transpositions or inversions of a segment of DNA could result in loss of *B8* expression without an alteration in band pattern. In any case, those *B8* loss mutants which seemed to have retained the B8 structural gene may offer an interesting opportunity to study control of HLA class I gene expression.

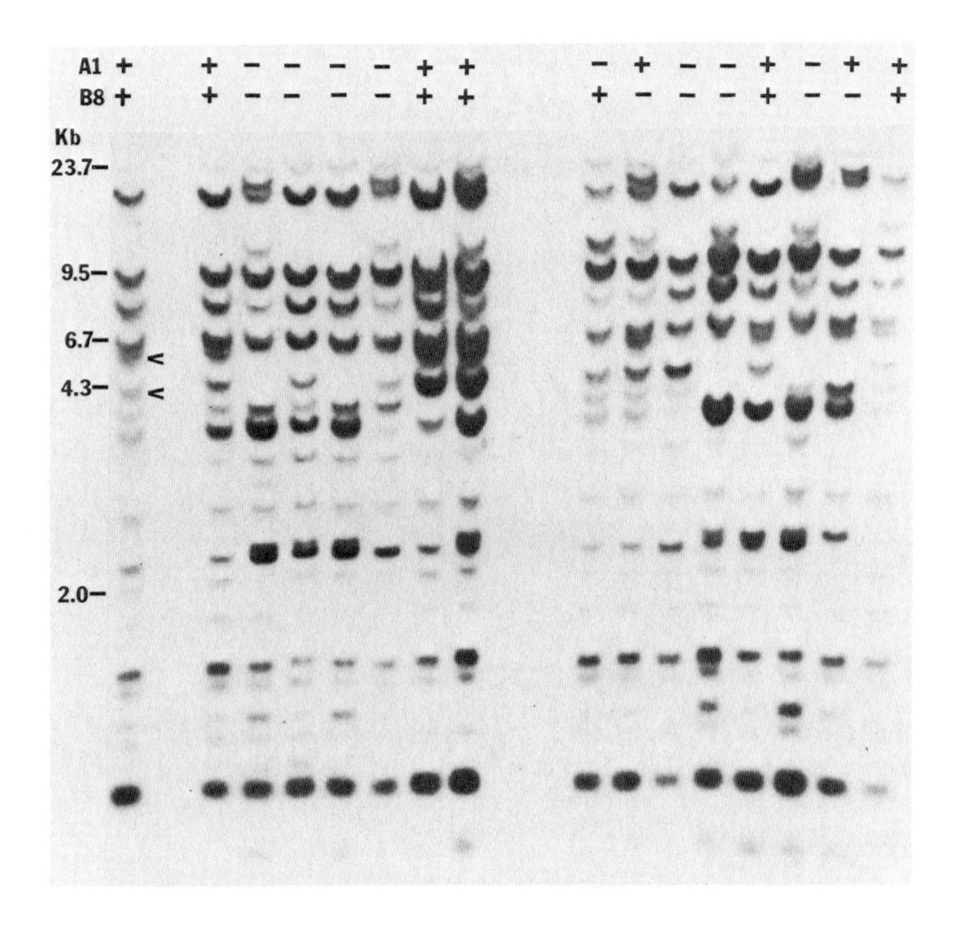

Fig. 6. Class I probe hybridization pattern of *PvuII* digested PBL DNA. The first lane has DNA from LCL 721. DNA in all other lanes was isolated from PBL's. The signs + and – depict the expression of *A1* and *B8* on LCL 721 and PBL's. Arrow heads indicate the 4.2 kb and 4.8 kb bands in LCL 721 that correlate with *B8* and *A1*, respectively.

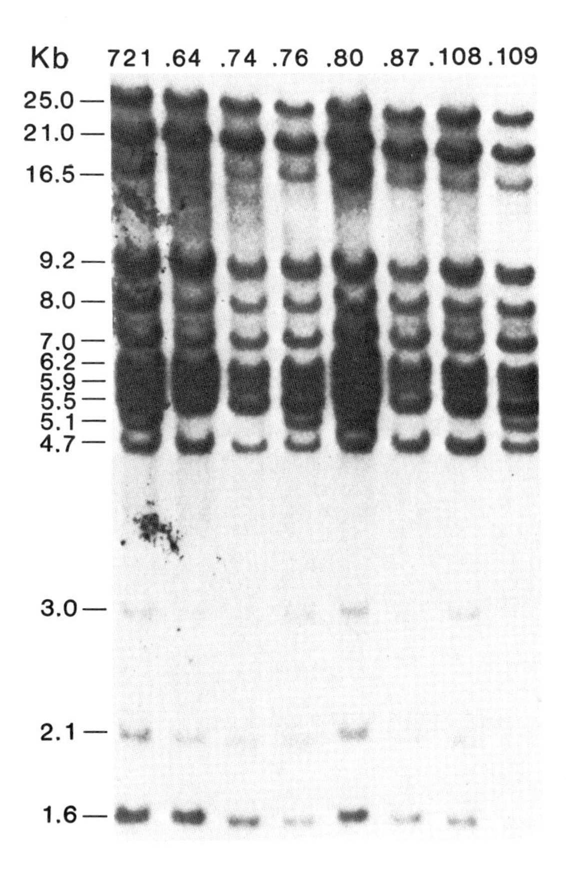

Fig. 7. Analysis of LCL 721 and seven HLA–A2 loss mutants. DNA was digested with *HindIII* and probed with the HLA–B7 cDNA clone.[12]

As described above, mutant 721.19 which lost the *GLO2; DR3; B8; A1*-containing haplotype, lost a 4.8 kb band in addition to the 4.2 kb band. Since mutants that lost expression of only *B8* lost only the 4.2 kb band, the 4.8 kb in the *B8*-containing haplotype contains a class I HLA gene other than *B8* (see below).

Correlation of HLA-B8 and HLA-A1 with PvuII fragments generated from DNA of normal lymphocytes

Peripheral blood lymphocytes (PBL) were purified from 15 individuals in order to determine if the *HLA-B8* and *HLA-A1* antigen-restriction fragment correlations detected in LCL 721 extended to the human population. Genomic DNA was purified from the PBL, digested with *PvuII*, and probed with the nearly full length HLA class I cDNA clone.[12] Figure 6 shows the hybridization patterns obtained compared to that from LCL 721.

The PBL panel contains seven $A1^+$ and eight $A1^-$ individuals. All seven of the $A1^+$ individuals have a hybridizing band at 4.8 kb while none of the eight $A1^-$ individuals have this band. Therefore, in this panel of 15 individuals the distribution of a class I hybridizing band at 4.8 kb correlates precisely with the expression of *A1*. Thus, a correlation that was observed among the LCL 721 mutants is strengthened by the PBL data.

The correlation of B8 expression with the 4.2 kb *PvuII* band in the PBL patterns is not as complete as that seen for the 4.8 kb fragment with *A1*. All six $B8^+$ individuals contained a 4.2 kb *PvuII* band that hybridized the nearly full length cDNA probe. However, this band was also found in four of the nine $B8^-$ individuals. Since the four $B8^-$ individuals having the 4.2 kb band do not share a detectable class I specificity, it is likely that several HLA alleles contain a 4.2 kb *PvuII* fragment.

Some additional features of the *PvuII* cut site polymorphism are evident from the panel of PBL DNA's. First, only one individual has a *PvuII* pattern that is identical to that of LCL 721 (Figure 6, compare lanes 1 and 2). All of the others have either lost or gained bands relative to LCL 721. The individual whose band pattern is identical to that of LCL 721 shares *A1, A2* and *B8* with LCL 721. Moreover, this individual does not express any HLA-A or HLA-B specificity not found in LCL 721, since B8 is the only HLA-B antigen typable in this person. It is interesting that such serological identity extends to the Southern blot patterns. However, determining the significance of this identity at the DNA level requires that more HLA identical individuals be compared by blot analysis.

Of the 20 *PvuII* fragments in LCL 721 that hybridized the nearly full length HLA cDNA probe, 14 are found in every individual analyzed. Yet, even in this limited population it is possible to detect several polymorphic *PvuII* fragments. It is interesting to speculate that in terms of restriction enzyme recognition site polymorphism HLA class I genes are divided into two sub-groups. One group, represented by the 14 *PvuII* fragments conserved between LCL 721 and the PBL panel might be derived from several class I genes which are highly conserved among alleles and loci. The second group represents those bands which vary between individuals and come from the 5′ region of class I genes. Probing with a larger variety of restriction endonucleases should permit evaluation of this speculation. Furthermore, our observations suggest that most of the polymorphic bands may be derived from the *HLA-A, HLA-B* and *HLA-C* genes. Comparison of the hybridization patterns obtained from several HLA identical and disparate individuals should be useful in determining the extent to which the HLA-A, HLA-B and HLA-C contribute to band variability.

Correlation of HLA-A2 expression with a HindIII generated fragment

Earlier work with loss mutants suggested that a *HindIII* fragment might be associated with expression of *A2*.[19] This was extended by probing *HindIII* digests of seven single A2 loss mutants with the Sood *et al*[12] cDNA clone. Compared to the parental LCL 721, four mutants no longer had a 5.1 kb hybridizing band (data unpublished). Thus, the earlier correlation of A2 with a *HindIII* generated fragment is substantiated. Explanations (see above) invoked to explain the failure to detect changes in *PvuII* generated band patterns of *B8* loss mutants apply as well to the three *A2* loss mutants with an apparent absence of *HindIII* generated bands. (Figure 7)

Use of HLA deletion mutants in determining the organization of class I-like genes

The correlation of DNA fragments generated by restriction endonucleases with known, genetically mapped HLA loci is one way of mapping the fragments. However, the number of bands which hybridized class I cDNA probes indicates that there are substantially more genes than can be accounted for by the *HLA-A, HLA-B* and *HLA-C* loci. HLA deletion mutants can be used to map such previously unknown HLA class I genes and, as described in DeMars *et al* (submitted for publication), MHC-region DNA that does not react with probes for known HLA genes.

Haplotype loss mutants 721.45.1 and 721.127 and the *HLA-A* null mutant 721.144 are particularly useful in this regard. In addition to lacking the 5.1 kb *HindIII* band, which closely correlates with *A2,* mutant 721.144 also lacks hybridizing *HindIII* generated bands at 15.0 kb, 7.0 kb, 5.9 kb and 2.1 kb and had reduced intensities at the 9.2 kb, 8.0 kb and 5.5 kb bands. Thus, mutant 721.144 has lost class I sequences linked to *A2*. Location of these additional class I genes relative to *A2* can be attempted by analyzing the hybridization pattern obtained from mutant 721.127, which has lost the entire *GLO1-A2-* containing haplotype (see ref. 39 for a detailed discussion). Thus, γ-ray-induced segregation between the *A2*-associated fragments generated by *HindIII* digestion of DNA locates previously undetected human class I-like genes. The evidence complements the γ-ray-induced segregations between *HLA-DR* and *HLA-SB.*[27,36] The portion of human chromosome No. 6 that contains histocompatibility and immune response genes may be much larger than is presently thought.[40,41]

It should be noted that the 9.2 kb, 7.0 kb and 2.1 kb bands hybridized the 5′ specific class I probe with less intensity than does the 5.1 kb band, which correlates with *A2.* Furthermore, the 9.2 kb and 7.0 kb bands hybridized the 3′ specific probe with an intensity comparable to the 5.1 kb band. The sequence homology between the *A2*-associated bands and *A2* in their coding sequence and the location of the *A2*-associated bands to the telomeric side of *A2* suggest that these bands might be the human analog of the murine TL region. In the mouse the TL loci map 1.2 centimorgans to the telomeric side of the *H-2D* locus.[37] By peptide mapping Tla antigens share some peptides (15-35%) with the H-2K, H-2D and H-2L antigens.[38]

Analysis of the *PvuII* DNA digests of mutants 721.45.1, 721.127 and 721. 144 allow additional *PvuII* fragments to be mapped (Figure 4). A-null mutant 721.144 lacks 4.8 kb and 4.2 kb bands because it was derived from 721.45.1,

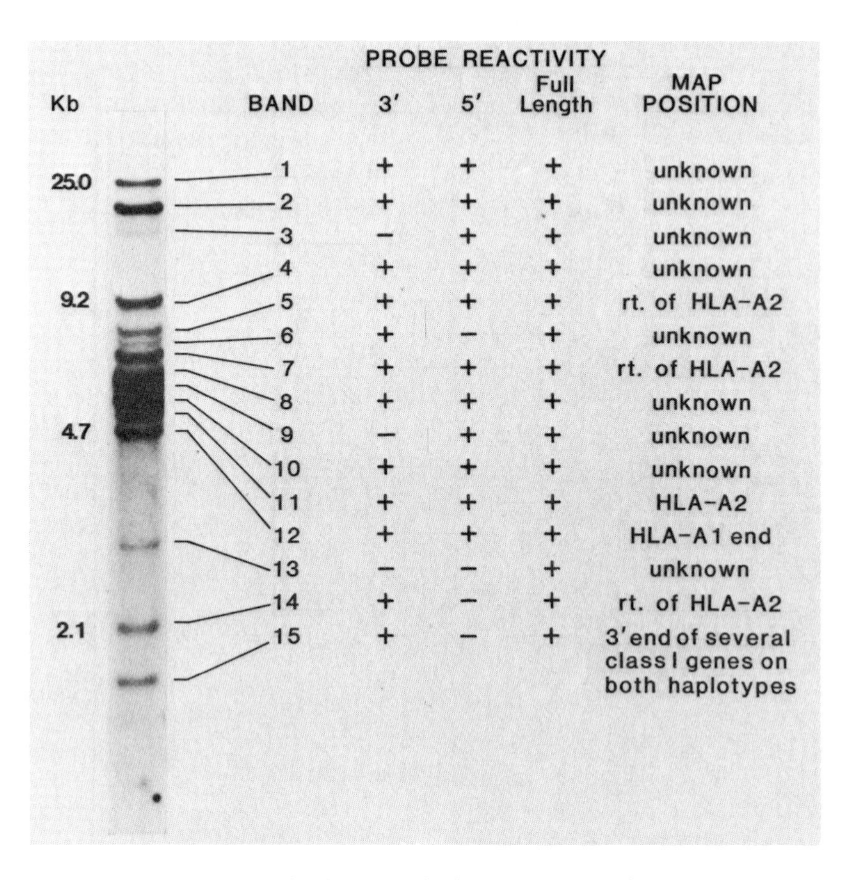

Fig. 8. *HindIII* Fragments of LCL 721. Probe reactivity and map position of the class I probe hybridizing *HindIII* generated fragments in LCL 721.

from which the entire *GLO2-B8*-containing haplotype was physically deleted. The analysis of 721.144, so far, indicates that it has a (proximal) break between *A2* and *B5* in the remaining haplotype. The other (distal) break in 721.144 must be on the telomeric side of the *A2* associated bands, which are distal to *A2*. Therefore, excepting the 4.8 kb and 4.2 kb bands, bands that are absent in 721.144 but present in 721.127, which has lost the *GLO1-B5*-containing haplotype, should map to the *HLA-A* locus (15 kb and 2.3 kb bands). Conversely, bands that are absent in 721.127 but present in 721.144 should map between the proximal breakpoint in 721.144 and the *B5* side of the haplotype (3.8 kb and 0.8 kb band). Bands that are absent in 721.127 and 721.144 but present in 721.45.1, which also lacks the *GLO2-B8*-containing haplotype, should map between the proximal breakpoint in 721.144 and the distal breakpoint in 721.127 (2.9 kb and 1.85 kb bands).

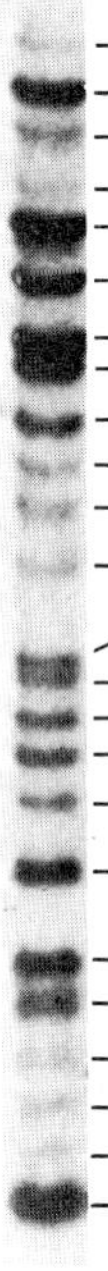

Band	kb	3′	5′	FL	Map Position
1	19.0	−	+	+	?
2	15.0	±	+	+	outside HLA-A2
3*	12.5	−	−	+	?
4	9.5	−	−	+	?
5*	7.6	±	+	+	?
6	6.2	±	+	+	?
7	5.0	−	+	+	?
8	4.8	−	+	+	HLA-A1
9	4.2	−	+	+	HLA-B8
10	3.8	−	+	+	HLA-B5
11	3.4	−	+	+	?
12	2.9	−	+	+	HLA-A2
13	2.3	−	+	+	?
14	2.25	+	−	+	outside HLA-A2
15	2.0	−	−	+	?
16	1.85	−	−	+	HLA-A2
17	1.68	+	−	+	?
18	1.35	−	−	+	?
19	1.05	+	−	+	?
20	0.95	−	−	+	?
21	0.80	−	−	+	HLA-B5
22	0.70	−	−	+	?
23	0.60	−	−	+	?
24	0.56	+	−	+	?

Fig. 9. Probe reactivity and map position of the class I probe hybridizing *PvuII* generated fragments in LCL 721.

Figures 8 and 9 summarize the analyses for the *HindIII* and *PvuII* digests, respectively. Clearly, the most striking results are with the *PvuII* pattern, where 30% of the class I hybridizing bands can be mapped within the HLA complex with the loss mutants we used. The isolation of a mutant analogous to 721.144 from the *DR3-A1* haplotype should be useful in mapping the remaining PvuII frgaments. Moreover, partial digests and the use of *PvuII* and *HindIII* in double digests of size selected *HindIII* and *PvuII* fractions, respectively, should also be useful in mapping additional bands. For example, the suggestion that the 18 kb *PvuII* band is from *A2* would be strengthened if this band were found to contain either the complete or a portion of the 5.1 kb *HindIII* band.

Complete elucidation of the HLA genomic structure will require "genomic walking" from one end of the MHC to the other. In man, such walking experiments must take into account the usual presence of two No. 6 chromosomes that are not identical with regard to HLA loci and restriction endonuclease cut sites. Reliable walking through the human MHC will be facilitated by the use of cells that express just one MHC haplotype *e.g.*, somatic cell hybrids that contain one human No. 6

or cells from apparently homozygous progeny of matings between close relatives. Mutants, such as 721.45.1, which have an entire haplotype physically deleted are alternative cells that can be used to unambiguously determine the nucleotide sequence of a specific haplotype. Furthermore, we have described how clones of DNA from mutants that have partial haplotype deletions can expedite genomic walking over long DNA intervals since, interstitial deletions join nucleotide sequences that may normally be widely separated (DeMars *et al,* submitted for publication).

SUMMARY

Mutants that had lost expressions of alleles of one or more HLA loci were isolated with immunoselection after γ-irradiation of human lymphoblastoid cell line LCL 721. DNA's from the mutants were digested with restriction endonucleases and analyzed by Southern blotting using probes for class I HLA genes. Eight polymorphic cut sites for *HindIII* and *PvuII* were discovered in class I-associated sequences of LCL 721. Losses of specific fragments generated by restriction enzymes could be associated with losses of specific antigenic expressions and it was possible in this way to assign *HLA-A1, HLA-A2* and *HLA-B8* to specific DNA fragments. Patterns of γ-ray induced segregations of DNA fragments permitted rough linkage alignment of about 30% of the fragments generated by *PvuII*. The resultant map showed that there are class I HLA genes on the telomeric side of the *HLA-A* locus.

ACKNOWLEDGMENTS

This work was supported in part by National Institutes of Health grants A1-15486, P30 CA 14520 and A1-18124 and by grants from the Searl Foundation and Leukemia Society to H.T. Orr. Also, we wish to thank C. Chang, R. Rudersdorf and G. Johnson for their technical assistance.

REFERENCES

1. Bach FH, van Rood JJ. The major histocompatibility complex - genetics and biology. *N Engl J Med* 1976; 295:806-13.
2. van Rood JJ, deVires RRP, Bradley BA. Genetics and biology of the HLA system. *In:* Dorf, ME, ed. The Role of the Major Histocompatibility Complex in Immunobiology. Garland, NY:STPM Press, 1981:59-114.
3. Orr HT, Lancet D, Robb RJ, Lopez de Castro JA, Strominger JL. The heavy chain of a human histocompatibility antigen (HLA-B7) contains an immunoglobulin-like region. *Nature* 1979a; 282:266-70.
4. Tragardh L, Rask L, Wiman K, Fohlman J, Peterson PA. Structure and sequence of an immunoglobulin-like HLA antigen heavy chain domain. *Proc Natl Acad Sci USA* 1979; 76:5839-42.

5. Coligan JE, Kindt TJ, Uehara H, Martinko J, Nathenson SG. Primary structure of a murine transplantation antigen. *Nature* 1981; 291:35-9.
6. Steinmetz M, *et al.* Three cDNA clones encoding mouse transplantation antigens: Homology to immunoglobulin genes. *Cell* 1981a; 24:125-34.
7. Orr HT, Lopez de Castro JA, Lancet D, Strominger JL. Complete amino acid sequence of a papain solubilized human histocompatibility antigen, HLA-B7. 2. Sequence determination and search for homologies. *Biochem* 1979; 18:5711-20.
8. Steinmetz M, *et al.* A pseudogene homologous to mouse transplantation antigens: Transplantation antigens are encoded by eight exons that correlate with protein domains. *Cell* 1981b; 25:683-92.
9. Malissen M, Malissen B, Jordan BR. Exon/intron organization and complete nucleotide sequence of an HLA gene. *Proc Natl Acad Sci USA* 1982; 79:893-97.
10. Evans GA, Margulies DH, Camerini-Otero RD, Ozato K, Seidman SG. Structure and expression of a mouse major histocompatibility antigen gene, *H-2L*^{d}. *Proc Natl Acad Sci USA* 1982; 79:1994-98.
11. Ploegh HL, Orr HT, Strominger JL. Molecular cloning of a human histocompatibility antigen cDNA fragment. *Proc Natl Acad Sci USA* 1980; 77: 6081-85.
12. Sood AK, Pereira D, Weissman SM. Isolation and partial nucleotide sequence of a cDNA clone for human histocompatibility antigen HLA-B by use of an oligodeoxynucleotide primer. *Proc Natl Acad Sci USA* 1981; 78:616-20.
13. Kvist S, *et al.* cDNA clone coding for part of a mouse H-2^d major histocompatibility antigen. *Proc Natl Acad Sci USA* 1981; 78:2772-76.
14. Moore KW, Sher BT, Sun YH, Eakle KA, Hood L. DNA sequence of a gene encoding a BALB/c mouse L^d transplantation antigen. *Science* 1982; 215:679-82.
15. Cosman D, Khoury G, Jay G. Three classes of mouse H-2 messenger RNA distinguished by analysis of cDNA clones. *Nature* 1982; 295:73-6.
16. Cami B, Bregegere F, Abastado JP, Kourilsky P. Multiple sequences related to classical histocompatibility antigens in the mouse genome. *Nature* 1981; 291:673-75.
17. Margulies DH, Evans GA, Flaherty L, Seidman JG. H-2 like genes in the Tla region of mouse chromosome 17. *Nature* 1982; 295:168-70.
18. Biro PA, Reddy VB, Sood A, Pereira D, Weissman SM. Isolation and analysis of human major histocompatibility complex genes. *In:* Walton AG, ed. Recombinant DNA. Amsterdam:Elsevier, 1981:41-9.
19. Orr HT, Bach FH, Ploegh HL, Strominger JL, Kavathas P, DeMars R. Use of HLA loss mutants to analyze the structure of the human major histocompatibility complex. *Nature* 1982; 296:454-56.
20. Goodenow RS, *et al.* Identification of a BALB/c *H-2L*^{d} gene by DNA-mediated gene transfer. *Science* 1982; 215:677-79.
21. Rajan TV. H-2 antigen variants in a cultured heterozygous mouse leukemia cell line. *Immunogenetics* 1977; 4:105-15.
22. Holtkamp B, Cramer M, Lemke H, Rajewsky K. Isolation of a cloned cell line expressing variant H-2^k using fluorescence-activated cell sorting. *Nature* 1981; 289:66-8.
23. Pious D, Hawley P, Forrest G. Isolation and characterization of HL-A variants in cultured human lymphoid cells. *Proc Natl Acad Sci USA* 1973; 70:1397-1400.
24. Andreotti P, Apgar J, Cresswell P. HLA-A2 as a target for cell-mediated lympholysis: Evidence from immunoselected HLA-A2 negative mutant cell lines. *Hum Immunol* 1980; 1:77-86.
25. Gladstone P, Fueresz L, Pious D. Gene dosage and gene expression in the HLA region: Evidence from deletion variants. *Proc Natl Acad Sci USA* 1982; 79:1235-39.

26. Kavathas P, Bach FH, DeMars R. Gamma-ray-induced loss of expression of HLA and glyoxalase I alleles in lymphoblastoid cells. *Proc Natl Acad Sci USA* 1980; 77:4251-55.
27. Kavathas P, DeMars R, Bach FH, Shaw S. SB: A new HLA-linked human histocompatibility gene defined using HLA-mutant cell lines. *Nature* 1981; 293:747-49.
28. Berger R, *et al.* Regional mapping of the HLA on the short arm of chromosome 6. *Clin Genet* 1979; 15:245-51.
29. Rigby PWJ, Dieckmann M, Rhodes C, Berg P. Labeling deoxyribonucleic acid to high specific activity *in vitro* by nick translation with DNA polymerase I. *J Mol Biol* 1977; 113:237-51.
30. Boyum A. Separation of leukocytes from blood and bone marrow. *Scand J Clin Invest* 1968; 21:77-89.
31. Blin H, Stafford DW. A general method for isolation of high molecular weight DNA from eukaryotes. *Nuc Acid Res* 1976; 3:2303-08.
32. Southern EM. Detection of specific sequences among DNA fragments separated by gel electrophoresis. *J Mol Biol* 1975; 98:503-17.
33. Wahl GM, Stern M, Stark GR. Efficient transfer of large DNA fragments from agarose gels to diazobenzyloxy methyl-paper and rapid hybridization by using detran sulfate. *Proc Natl Acad Sci USA* 1979; 76:3683-87.
34. Abrahamson S, Wolff S. Reanalysis of radiation-induced specific locus mutations in the mouse. *Nature* 1976; 264:715-19.
35. Christensen R. Specificity and frequency of ultraviolet-induced re-version of an iso-1-cytochrome C ochre mutant in radiation-sensitive strains of yeast. *J Mol Biol* 1974; 85:137-62.
36. Shaw S, Kavathas P, Pollack MS, Charmot D, Mawas C. Family studies define a new histocompatibility locus, SB, between HLA-DR and GLO. *Nature* 1981; 293:745-47.
37. Flaherty L. The Tla region antigens. *In:* Dorf ME, ed. The Role of the Major Histocompatibility Complex in Immunobiology. New York:Garland STPM Press, 1980:33-57.
38. Yokoyama K, Stockert E, Old LJ, Nathenson SG. Structural comparisons of TL antigens derived from normal and leukemia cells of TL+ and TL− strains and relationship to genetically linked H-2 major histocompatibility complex products. *Proc Natl Acad Sci USA* 1981; 78:7078-82.
39. Orr HT, DeMars R. Class I-like HLA genes map telomeric to the HLA-A2 locus in human cells. *Nature* 1983; 302:534-536.
40. DeMars R, Chang CC, Rudersdorf RA. Dissection of the D-region of the major histocompatibility complex by means of induced mutations in a lymphoblastoid cell line. *Human Immunol.* (in press)
41. DeMars R, *et al.* Dissociation in expression of MBI/MTI and DRI alloantigens in mutants of lymphoblastoid cell nine. *J. Immunol.* (in press)

DNA POLYMORPHISMS IN THE B–GLOBIN CLUSTER: A STRATEGY FOR DISCOVERING NEW MUTATIONS

Haig H. Kazazian, Jr.
Stylianos E. Antonarakis
Tu-chen Cheng
Corinne D. Boehm
Pamela G. Waber

INTRODUCTION

Since Kan and Dozy (1978) discovered the first known DNA polymorphism in man in the β-globin cluster, a large number of other polymorphisms have been discovered in this cluster.[1,3-6] We have studied these DNA polymorphisms in various ethnic groups in an effort to obtain both basic knowledge and information that could be applied to clinical medicine.

We have found that a small number of common DNA sequences specify the 5′ region of the β-globin gene cluster and that a small number of common sequences specify the 3′ region of this cluster. Between these two regions is a potential hotspot for recombination. In addition, we have observed that generally different mutations leading to a single-gene disease, β-thalassemia, have occurred on different normal chromosome backgrounds. This has led us to propose a strategy which is potentially applicable for molecular characterization of other heterogeneous single-gene disorders.

RESULTS AND DISCUSSION

Polymorphic Sites in the β-Globin Gene Cluster

The β-globin gene cluster occupies about 50 kb of DNA in the short arm of chromosome 11 in man.[2] Embryonic (ϵ), fetal ($^{G}\gamma$ and $^{A}\gamma$), and adult (δ and β) globin genes are arranged in a 5′ to 3′ (left to right) order and the gene sequences account for about 15% of the DNA in this region (Fig. 1). In Figure 1, $\psi\beta_1$ is the β-like nonfunctional gene (pseudogene). The function of the flanking sequences, which contain certain repeated DNA regions, is not well understood.

ISBN 0-12-492220-1

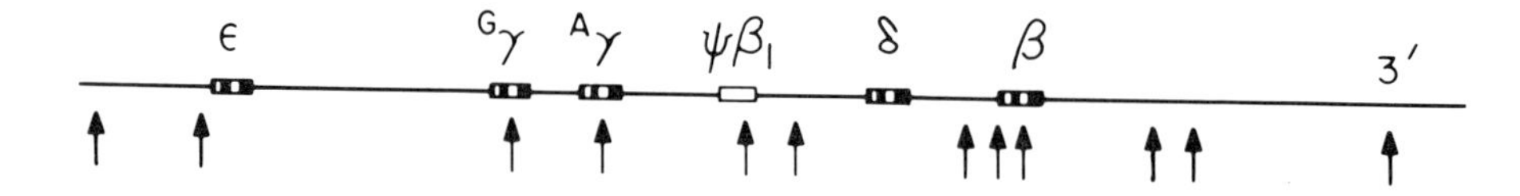

Fig. 1. β-globin gene cluster and the location of twelve known polymorphic restriction sites (shown by arrows). ε = embryonic gene. $^G\gamma$, $^A\gamma$ = fetal genes. $\psi\beta_1$ = pseudogene. δ, β = adult genes.

Using the methods of restriction endonuclease analysis of DNA obtained from peripheral leukocytes, a number of DNA polymorphisms have been found in the β-globin gene cluster.[3-6] In this discussion we will concentrate on sites which are polymorphic to the extent that the less common variant (whether presence or absence of the site) is present in one or more racial groups with a frequency greater than 5%. At present using this definition 12 polymorphic sites have been found in the β-gene cluster (Fig. 1). Nine of these sites are public (occur in all racial groups) and the frequency of the less common allele is greater than 5% in all racial groups studied to date. The remaining polymorphic sites, *TaqI, HinfI,* and *HpaI,* are private in that they are polymorphic in Blacks but not in other racial groups. It is of interest that certain private polymorphisms such as *TaqI* and *HinfI* have attained quite high frequencies (25-35%) in Blacks even though they they are very rare in other racial groups (unpublished data). All polymorphisms on which data are available (six of the 12) are single nucleotide substitutions. The polymorphisms in the β-globin gene cluster are in the following locations: Seven of the 12 are in flanking DNA, three are in intervening sequences, one is in a pseudogene sequence, and one is in the coding region of the β gene.

Study of restriction site polymorphisms has also led to the elucidation of nucleotide polymorphisms which are not detectable by restriction site analysis. In the course of sequencing various mutant genes (especially β-thalassemia) whose neighboring DNA contained different patterns of polymorphic restriction sites, nucleotide polymorphisms were found in the β gene (see below).

Common Sequence Variation in the β-Globin Gene Cluster

Over the past 18 months we have studied the polymorphic restriction endonuclease sites of 200 normal and 200 mutant β-globin gene clusters, and correlated the information with sequence analysis of a subset of over 20 β-globin genes.[7] These data have allowed us to make some general conclusions about the sequence variation within this cluster and, by analogy, that within other gene clusters.

The general picture which emerges is that there exists a relatively small number of common haplotypes with respect to restriction site pattern (Fig. 2).

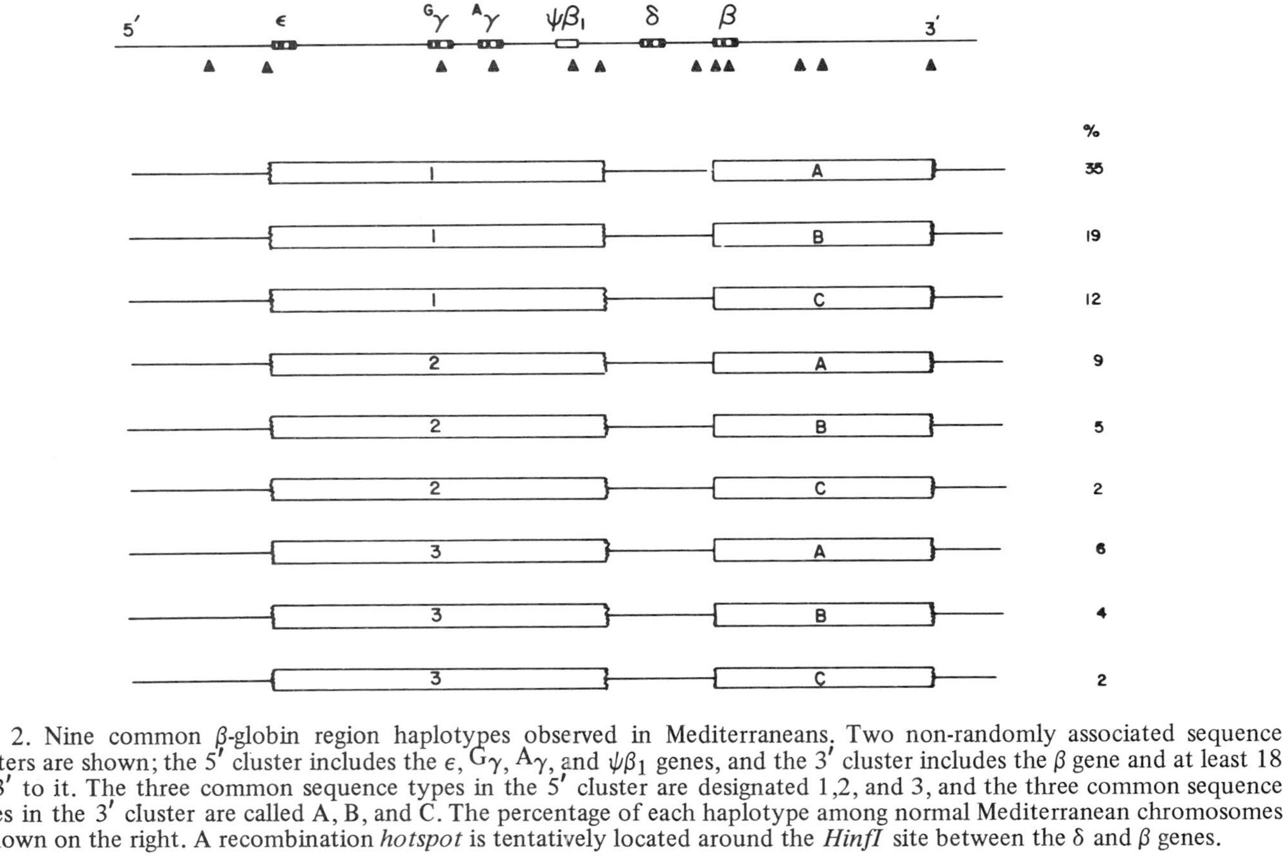

Fig. 2. Nine common β-globin region haplotypes observed in Mediterraneans. Two non-randomly associated sequence clusters are shown; the 5′ cluster includes the ε, $^G\gamma$, $^A\gamma$, and $\psi\beta_1$ genes, and the 3′ cluster includes the β gene and at least 18 kb 3′ to it. The three common sequence types in the 5′ cluster are designated 1,2, and 3, and the three common sequence types in the 3′ cluster are called A, B, and C. The percentage of each haplotype among normal Mediterranean chromosomes is shown on the right. A recombination *hotspot* is tentatively located around the *HinfI* site between the δ and β genes.

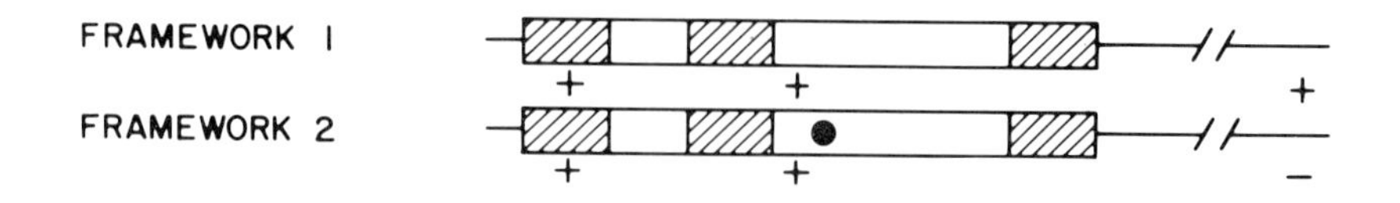

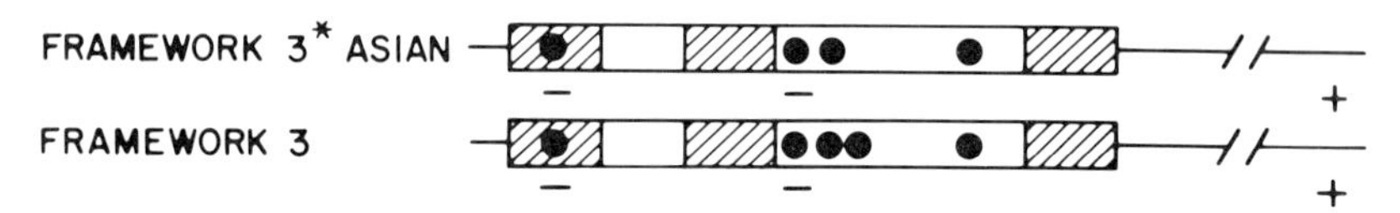

Fig. 3. Normal β-globin gene frameworks. Filled areas are coding regions in the β gene and open areas are intervening sequences. Frameworks 1, 2, and 3 are found in Mediterraneans, and their patterns of polymorphic restriction sites shown as +++, ++−, −−+ correspond to 3′ sequence types A, B, and C of Figure 2, respectively. Black dots in frameworks 2 and 3 indicate the locations of single nucleotide differences from the sequence of framework 1. Framework 3* Asian is found only in Asians and lacks one of the nucleotide substitutions found in framework 3. The space between framework 2 and framework 3* Asian is meant to suggest the possible existence of evolutionary intermediates.

A haplotype is a particular pattern of restriction sites in the DNA region studied. The DNA region studied covers the entire β-globin gene cluster from the embryonic ϵ gene to the 3′ region of the β-globin gene. Each of these common haplotypes appears to represent a common DNA sequence. In the β-globin gene (Fig. 3) and extending 18 kb 3′ to the gene there are three common sequence types (β-gene frameworks 1, 2, and 3) in the Mediterraneans and three common sequence types (β gene frameworks 1, 2, and 3* Asian) in Asians. Two of these common sequence types are present in both groups, while one varies by a single nucleotide between the groups. B-gene frameworks 1 and 2 differ by a single nucleotide in IVS-2 while framework 3 in Mediterraneans has the nucleotide polymorphism of framework 2 and 4 additional nucleotide polymorphisms. Three of these additional polymorphisms are in IVS-2 and one is a silent substitution in codon 2. In Asians framework 3* lacks one of the IVS-2 changes found in framework 3 of Mediterraneans. By restriction site analysis, Blacks also appear to have the three β-gene frameworks, but whether or not minor sequence differences exist between the Black frameworks and those of Mediterraneans and Asians has not yet been established.

On the 5′ side of the β gene spanning at least 32 kb from the ϵ gene to the 3′ side of the $\psi\beta_1$ gene there are also three common sequence types (Fig. 2)[6] These common sequence types are also present in both Mediterraneans and Asians,

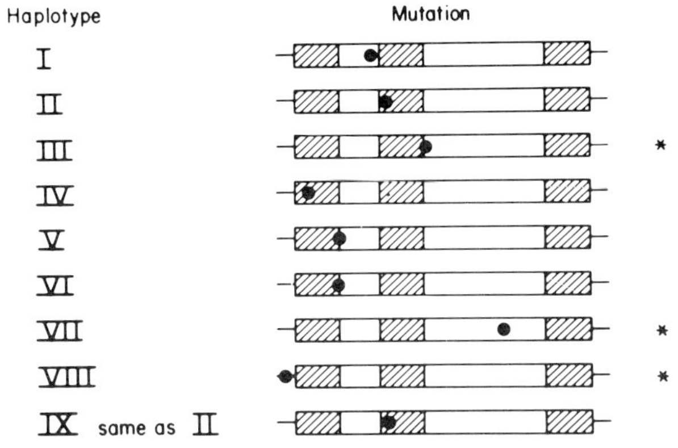

Haplotype Mutation

I

II

III *

IV

V

VI

VII *

VIII *

IX same as II

Fig. 4. β-thalassemia mutations associated with different haplotypes in Mediterraneans. Nine different haplotypes for seven polymorphic restriction sites were found associated with β^{thal} chromosomes. At least one β gene associated with each haplotype was analysed, and a different mutation was associated with eight of the haplotypes. Haplotype IX had the same mutation as haplotype II.[7] Mutation sites are shown as black dots. Asterisks at the right of haplotypes III, VII, and VIII, indicate that these mutations can be detected directly by restriction site analysis.

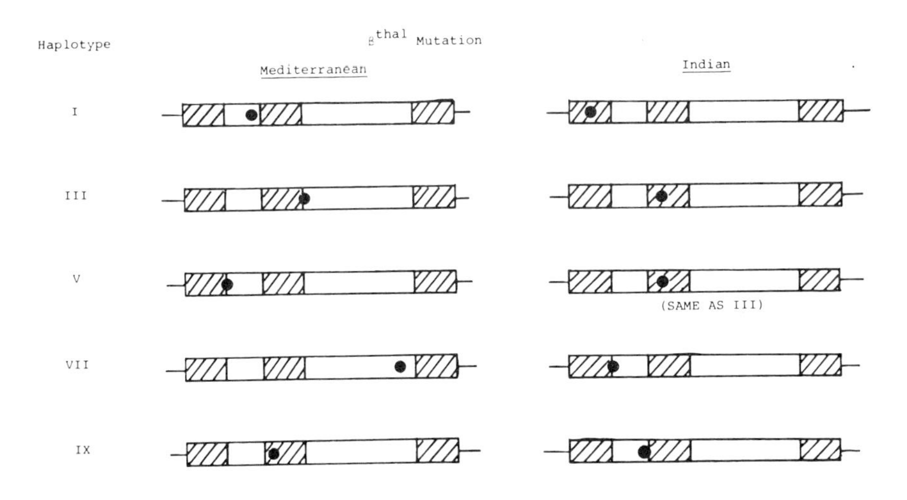

Fig. 5. Mutations producing β-thalassemia are different from one ethnic group to another, even when chromosomes with the same haplotype are studied. Haplotypes I, III, V, VII, and IX are present in both Mediterraneans and Asian Indians. The mutations present in Indians are different from those in Mediterraneans.

but Blacks have different common sequence types. Between these two regions in the β gene cluster: (a) the gene and its 3′ side and, (b) the ϵ, ${}^{G}\gamma$, ${}^{A}\gamma$, $\psi\beta_1$ genes, there is a DNA segment of 11 kb within which randomization of the 5′ and 3′ regions seems to occur (Fig. 2). This segment is thought to be a relatively high recombination area, and a possible *recombination hotspot* is now believed to be located around a polymorphic *HinfI* site 1 kb 5′ to the β gene. Since the common sequence types of the 5′ and 3′ regions are combined in a relatively random fashion, recombination within the β-globin gene cluster may be restricted to this region.

TABLE 1. Point Mutations Producing β-Thalassemia

Ethnic Group	β^{thal} Genes Anal.	Muts. Found	β^{thal} Chromos. Anal.	Different Haplos. Found	Expected Total No. Muts.
Mediterraneans	30	11	100	10	14
Asian Indians	6	5	36	10	10
Chinese	1	1	10	3	10
American Blacks	1	1	10	4	10
Kurdish Jews	2	2	8	3	4
Southeast Asians	2	1	2	2	5
North Africans					3
Others					5
Total		21			~60

The number of β-thalassemia mutations (~60) is estimated from: (1) the association of different mutations with different haplotypes, (2) the fact that different mutations are found in each ethnic group, and (3) the large number of haplotypes yet to be found in certain poorly-studied ethnic groups, *e.g.* Chinese and American Blacks.

β-Thalassemia Mutations and the Normal Sequence Variation in the β-Globin Gene Cluster

The number of different β^{thal} alleles characterized has been greatly expanded by a new strategy. This strategy depends upon determining the haplotype for each chromosome carrying a β-thalassemia gene by using restriction endonuclease analysis (Fig. 4). One can then sequence only those mutant genes present in different haplotypes. This strategy has produced a high yield of previously undiscovered mutations in Mediterraneans.[7] Moreover, mutations in one ethnic group do not appear to overlap with those present in another group. When the haplotypes associated with β-thalassemia in Mediterraneans and Asian Indians are compared, five are common to both groups. Four different point mutations have been characterized associated with these haplotypes in Asian Indians, and these are different from those found with the same haplotypes in Mediterraneans (Fig.5). In addition, a fifth mutation has been characterized in an Asian Indian which is not found in Mediterraneans. Up to now the β-thalassemia alleles have been studied in detail only in these two groups. After Chinese, Black Africans, and other groups with a high incidence of this disease are studied, we expect that the number of different β^{thal} alleles may be 60 or more (Table 1). Of course, many different kinds of mutations could reduce gene expression and produce β-thalassemia. The kinds of mutations observed to date and their functional characteristics are discussed by Orkin in his chapter. A new mutation of some interest was discovered because it obliterated the same *Mst* II site which is obliterated by the sickle cell β-globin allele. This mutation is a simple deletion of the nucleotide that undergoes substitution in the β^S gene. The normal β gene at codon 6 is GAG, and the β^S gene is GTG. This β-thalassemia gene is GG, a frameshift mutation.

Our sequencing strategy has relied on two premises: (1) that different haplotypes are associated with different β-thalassemia mutations, and (2) that a small number of mutations, one or two, make up the bulk of mutations associated with any one haplotype. Thus, it is important that we determine the extent of association of mutation with haplotype. To date, our data continue to support a strong association and they suggest that our sequencing strategy is a reasonable one.

Thirteen β^{thal} alleles associated with haplotype I chromosomes have been studied, four by direct DNA sequencing and nine by S_1 mapping of erythroid RNA. The IVS-1 mutation described by Spritz *et al.*[8] was found on 12 of the 13 chromosomes. Likewise, all eight mutations associated with haplotype III examined to date (one by DNA sequencing, the rest by restriction analysis of genomic DNA and Northern blot analysis of erythroid RNA) contain the same IVS-2 splice junction mutation.[9] Other examples could be cited, but all support the contention that only one or two mutations account for the bulk of mutations associated with each haplotype. This suggests that each β^{thal} mutation has expanded greatly in frequency in the population in which it originated.

In general, β-gene mutations appear to be more recent events than the sequence variation associated with the observed DNA polymorphisms. However, it is possible for them to lose their association with the sequence 5′ to the β-gene by recombination. The fact that the same mutation is associated with one β-gene framework and two different 5′ sequence types suggests that it originated from a single mutation and subsequently was recombined with a different 5′ sequence type.

On the other hand, discovery of a β-gene mutation on two or more different β-gene frameworks suggests that the mutation may have either had multiple independent origins or migrated from one framework to another by some form of interallelic gene conversion mechanism.[10] Indeed, the β^E mutation has been found on two different β-gene frameworks[11] and recently the mutation at codon 39 producing β-thalassemia has been found on two different β-gene frameworks in Italians (Fig. 6.).

Implications for Characterization of Genetic Disease

This description of the variation in common sequence types and in recombination rates within a gene cluster has implications for the study of genetic diseases. In order to provide a *complete* description of the molecular basis of a genetic disease the following plan would be useful, given our experience with β-thalassemia and hemoglobinopathies:

1. Obtain genomic clones containing the normal gene in question and neighboring sequences to serve as molecular probes.
2. Use these probes to discover DNA polymorphisms within and adjacent to the gene and by family studies discover the non-random associations such as those with only three combinations for two or three restriction site polymorphisms.
3. Study genomic DNA of ten or more patients *homozygous* for the disease in question and their family from each affected ethnic group to discover the mutant gene on various chromosome backgrounds.
4. Sequence one mutant gene associated with each of these backgrounds to discover different mutant alleles.

One might expect that each chromosome background of each ethnic group would be associated with one or more different mutant alleles. If there exist many possible mutations which eliminate protein function or synthesis, and positive selection for heterozygotes is weak (as is expected for most mutations), then many different mutant alleles should be found and nearly all *clinical homozygotes* may be genetic compounds.

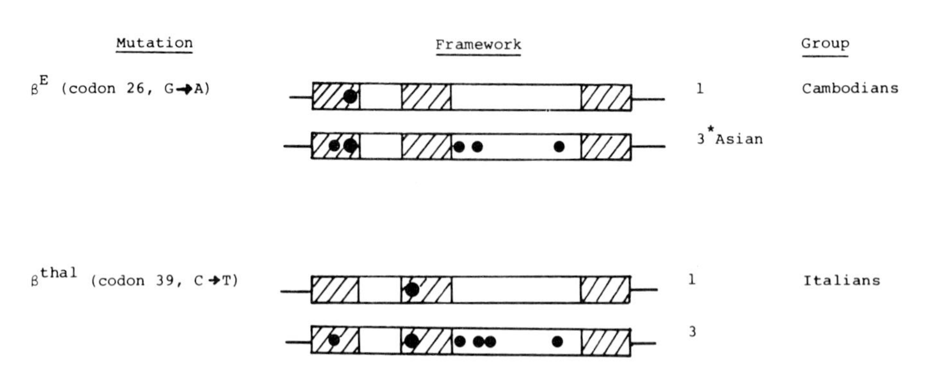

Fig. 6. Two examples of the same mutation in two different β-gene frameworks. Large dots are the mutations β^E (codon 26, G→A) and β^{thal} (codon 39, C→T) and small dots are common nucleotide polymorphisms.

SUMMARY

We have discussed two major uses of DNA polymorphisms in genetic research and clinical medicine. First, they have been of great value in defining a small number of common sequences in the β-globin gene cluster and regions within which recombination may be restricted. Second, they have led to a screening procedure which not only has been of great value for the molecular characterization of β-thalassemia mutations but also has implications for the characterization of other single gene disorders.

ACKNOWLEDGMENTS

Haig H. Kazazian, Jr. is supported by grants from the National Institutes of Health and the March Dimes. Tu-chen Cheng is supported by a grant from the Nationat Institutes of Health.

Some material has been reprinted from "DNA polymorphisms in β-globin gene cluster: Use in discovery of mutations and prenatal diagnosis. *In:* Recombinant DNA Application to Human Disease, Caskey CT, White R (eds.). Banbury Reports, No. 10 (in press).

REFERENCES

1. Jeffreys A. DNA sequence variants in the $^{G}\gamma$, $^{A}\gamma$, δ, β-globin genes of man. *Cell* 1979; 18: 1-10.
2. Efstratiadis A, *et al.* The structure and evolution of the human β-globin gene family. *Cell* 1980: 21:653.
3. Kan YW, Dozy AM. Polymorphism of DNA sequence adjacent to the human β-globin structural gene: relationship to sickle cell mutation. *Proc Natl Acad Sci* 1978; 75:5631-35.
4. Tuan D, Biro PA, de Riel JK, Lazarus H, Forget BG. Restriction endonuclease mapping of the human β-globin gene loci. *Nucl Acids Res* 1979; 6:2519.
5. Kan YW, Lee KY, Furbetta M, Anguis A, Cao A. Polymorphism of DNA sequence in the β-globin gene region. *New Eng J Med* 1980; 302:185-88.
6. Antonarakis SE, Boehm CD, Giardina PJV, Kazazian HH, Jr. Nonrandom association of polymorphic restriction sites in the β-globin gene cluster. *Proc Natl Aca Sci* 1982; 79: 137-41.
7. Orkin SH, *et al.* Linkage of β-thalassemia mutations and β-globin gene polymorphisms with DNA polymorphisms in the human β-globin gene cluster. *Nature* 1982; 296:627-31.
8. Spritz RA, *et al.* Base substitution in an intervening sequence of a β^{+}-thalassemic human globin gene. *Proc Natl Acad Sci* 1981; 78: 2455-59.
9. Treisman R, Proudfoot N, Shander M, Maniatis T. A single-base change at a splice site in a β^{o}-thalassemia gene causes abnormal RNA splicing. *Cell* 1982; 29:903-11.
10. Dover G. Molecular drive: A cohesive model of species evolution. *Nature* 1982; 299: 111-17.
11. Antonarakis SE, *et al.* Evidence for multiple origins of the β^{E}-globin gene in South East Asia. *Proc Natl Aca Sci* 1982; 79:6608-11.

MOLECULAR ANALYSIS OF A HUMAN DISEASE: B–THALASSEMIA

Stuart H. Orkin

B-thalassemias are inherited disorders of a β-globin synthesis that result from mutations within the β-globin gene. Elucidation of the precise molecular defects associated with these syndromes defines regions of the gene that are critical in complex processes such as RNA processing and gene transcription. In addition, full characterization of the defects that lead to these severely debilitating conditions may appreciably contribute to improved prenatal detection of specific mutant genes. Over the past few years the application of recombinant DNA methods to the analysis of β-thalassemia has led to dramatic advances in our understanding of the molecular pathology. Some of these aspects will be reviewed here.

SYSTEMATIC SEARCH FOR NEW GENE MUTANTS

Complete dissection of the molecular basis of an inherited disease requires a systematic approach to the identification of the existing alleles in a population. Clinically, β-thalassemia is a heterogeneous entity, suggesting that many different gene mutations exist. In order to avoid bias in favor of the most common mutant alleles we have developed a systematic approach to the study of this disease that combines analysis of restriction site polymorphisms with molecular cloning of selected β-globin genes.[1] The various restriction enzyme cleavage sites that are known to be polymorphic within the β-globin cluster can be used to construct a haplotype for this chromosomal region in uncloned DNA samples.[2] The nature of the sites chosen and the description of the observed haplotypes are summarized by Kazazian in the accompanying paper in this volume. Among a typical population group, such as Mediterranean β-thalassemics, only a limited number of different haplotypes are observed.[1,2] It seemed likely to us, and is borne out by experimental analysis, that each haplotype might reflect a different allele of β-thalassemia. If this were the case, cloning of representative members of each haplotype in a population would constitute a highly efficient means of nearly completely dissecting the entire group. With some modifications, the simple correlation of a haplotype with a specific gene mutation has been observed. For example,

ISBN 0-12-492220-1

TABLE 1. General Approach to Systematic Search for New Globin Gene Mutants

1. Characterize DNA's by polymorphism haplotypes.
2. Clone representative members of each haplotype.
3. Examine DNA sequence for putative thalassemia mutations and polymorphisms within the gene.
4. Express mutant genes in heterologous cell system to prove nature of putative thalassemia mutations.

among Mediterranean β-thalassemics we initially observed nine haplotypes and succeeded in characterizing eight different β-thalassemia genes.[1] DNA sequencing of cloned mutant β-globin genes provides the most direct method for identification of putative pathologic mutations as well as common DNA sequence polymorphisms within the β-gene.[1] Expression of the cloned genes introduced into heterologous cells allows definitive association of particular nucleotide sequence changes with phenotypic effects on gene expression. The strategy to the analysis of β-thalassemia is outlined in Table 1.

Although several different gene expression systems have been utilized for functional study of cloned genes,[3,4] investigation of globin gene transcription and RNA processing is most easily accomplished in a transient expression system such as that developed by Trieisman *et al*[5] and summarized in Figure 1. In this system the cloned β-globin gene is introduced in HeLa cells linked to plasmid and SV40 DNA sequences *via* transfection. Although several vector constructs have been employed, high level proper transcription of the β-globin gene requires the presence of enhancer sequences of SV40 *in cis* to the globin sequences.[3,5] After 1–2 days of incubation of the HeLa cells following transfection sufficient RNA can be obtained to determine the abundance and structure of globin RNA transcripts produced from either the normal or mutant genes.

MUTATIONS LEADING TO B–THALASSEMIA

At present 20 specific defects have been defined in β-thalassemia. These are summarized in Table 2. The mutations are most easily classified by the phase of gene expression that is adversely affected. In principle, we may divide them into mutations disturbing mRNA function and stability, RNA processing, and gene transcription. Nucleotide sequence alterations that generate either nonsense or frameshift mutations in the β-gene produce β^0-thalassemia, the absence of β-globin synthesis. For reasons that are not entirely clear all such nonfunctional mRNA's appear unstable *in vivo*, although to varying degrees. Although a relatively common cause of β-thalassemia, these mutations do not provide special insights into aspects of gene expression and will not be considered further here. Instead, I would like to describe novel findings regarding RNA processing and transcription derived from studies of thalassemic genes.

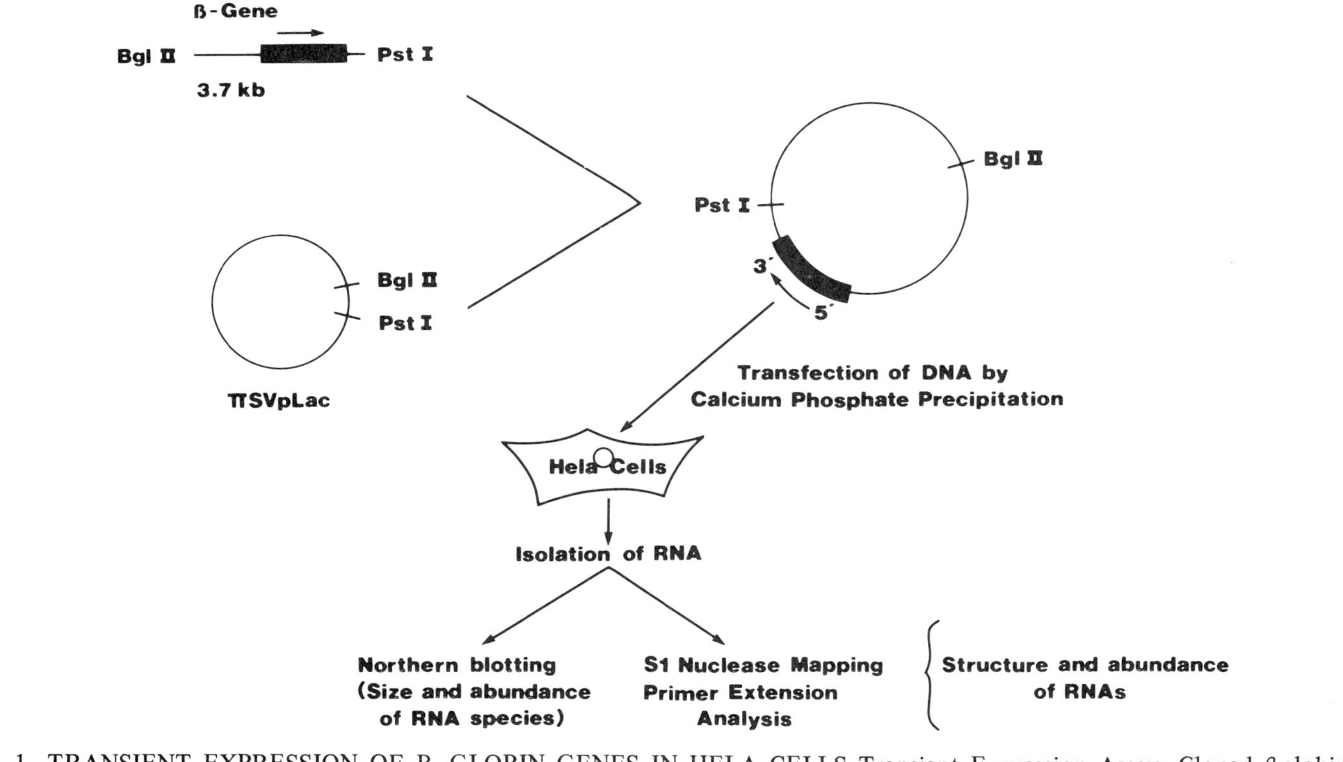

Fig. 1. TRANSIENT EXPRESSION OF B–GLOBIN GENES IN HELA CELLS Transient Expression Assay. Cloned β-globin genes, contained in a 3.7 kb *BglII-PstI* fragment, are subcloned into πvx expression plasmid developed by Treisman *et al.*[5] The globin recombinant is introduced into HeLa cells by calcium phosphate precipitation (transfection); 24–48 hours later RNA is isolated and assayed for globin RNA content and structure by Northern blotting, S1 nuclease mapping, and primer extension analysis.

TABLE 2. Mutations in B-Thalassemia

Mutation Class	Type	Ethnic Origin	References
I. Nonfunctional mRNA			
a) Nonsense mutations:			
codon 17	β^0	Chinese	30
codon 39	β^0	Mediterranean	31,32,33
b) Frameshift mutations:			
-2 codon 8	β^0	Turkish	31
-1 codon 18	β^0	Indian	unpublished
-1 codon 44	β^0	Kurdish	34
-4 codons 41/42	β^0	Indian	unpublished
+1 codons 8/9	β^0	Indian	unpublished
-1 codon 8	β^0	Mediterranean	unpublished
II. RNA Processing Mutants			
a) Splice junction substitutions:			
IVS-1 position 1 G→A	β^0	Mediterranean	1
IVS-2 position 1 G→A	β^0	Mediterranean	5
IVS-1 3′-end 25 nt deletion	β^0	Indian	unpublished
b) Consensus substitutions:			
IVS-1 position 5 G→C	β^+	Indian	7,unpublished
IVS-1 position 6 T→C	β^+	Mediterranean	1
c) Internal IVS substitutions:			
IVS-1 position 110	β^+	Mediterranean	8,9
IVS-2 position 705	β^+	Mediterranean	12
IVS-2 position 745	β^+	Mediterranean	1
d) Coding region substitutions affecting RNA processing:			
codon 26 G→A	β^E	Southeast Asian	16
codon 24 T→A	β^+	Black	17
III. Transcriptional Mutants			
-87 C→G	?(prob. β^+)	Mediterranean	1
-28	β^+ (prob.)	Kurdish	22

MUTATIONS LEADING TO ABNORMAL RNA PROCESSING

When it became apparent that intervening sequences were a common feature of globin and other eucaryotic genes, it was proposed that some forms of thalassemia might result from abnormalities in RNA processing.[6] Indeed, this has turned out to be the case. For purposes of discussion, however, it is useful to subdivide these processing mutants into four classes.

Comparisons of many normal gene sequences have led to formulation of consensus sequences for coding region-intervening sequence junctions.[35] Although somewhat flexible in their sequences, the donor or 5′ splice site always has a GT dinuecleotide, and the 3′ splice site an AG dinucleotide. Neighboring these invariant dinucleotides are preferred sequences that vary among the different normal genes that have been reported. Two β-thalassemic genes have been found in which the invariant donor GT is changed to AT, either at the splice site of IVS-1 or IVS-2.[1,5] In both instances, the nucleotide sequence alteration prevents normal RNA processing at the 5′ end of the intervening sequence and leads to alternative RNA processing at other donor-like sequences either within coding or intervening sequences in the precursor RNA transcript.[5,7] These mutations are therefore of the β^0-type.

A second class of processing mutation includes nucleotide changes in the consensus regions but not within the invariant nucleotides. Two such mutations have been identified within the consensus sequence of the IVS-1 donor site[1] (unpublished). Both mutations lead to decreased splicing at the normal donor site and alternative RNA splicing that utilizes the same donor-like sequences employed in transcripts from the GT-AT mutant gene.[7] The splicing patterns of these various mutations in the donor sequence of IVS-1 are summarized in Figure 2. In general, the severity of impairment in globin RNA production in HeLa cells mirrors the relative severity of these genes *in vivo*. Since some normal globin RNA is made from the mutant genes with substitutions outside the invariant dinucleotide, they are of the β^+-type, that is some normal β-globin is directed by the defective allele. These latter splicing mutants emphasize the complexity of RNA processing. Seemingly trivial nucleotide sequence changes within the consensus sequence produce significant effects on both the efficiency and accuracy of RNA processing.

The third class of processing mutants includes those in which a nucleotide substitution with an intervening sequence generates a new splicing signal. Three mutations of this kind are known, as listed in Table 2. In the first characterized gene of this type, the common β^+-thalassemia gene among Mediterraneans, a G→A change at position 110 of IVS-1 produces an internal acceptor-like sequence that is used in RNA processing more efficiently than the normal site at the end of IVS-1.[8-11] In this manner, the majority of RNA processed from this gene is abnormal and contains an extra 19 nucleotides derived from the intervening sequence. Again, since some normal globin RNA is formed, the defect is of the β^+-type. Two

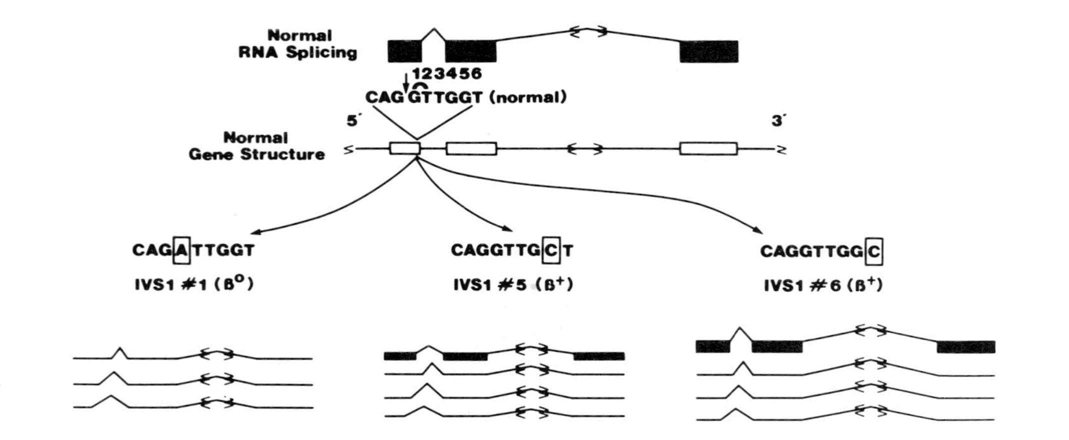

Fig. 2. Nucleotide Substitutions of the 5′-IVS-1 Consensus Sequence. The DNA region mutated in three β-thalassemic genes are depicted. The specific nucleotide changes are boxed. The normal RNA splicing pattern is shown at the top. Below, the splicing patterns for the mutants are shown. The height of the coding boxes roughly approximates the level of steady-state RNA seen in HeLa cells.

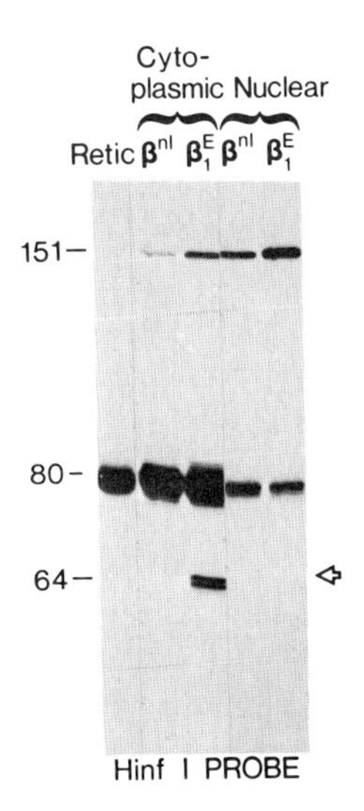

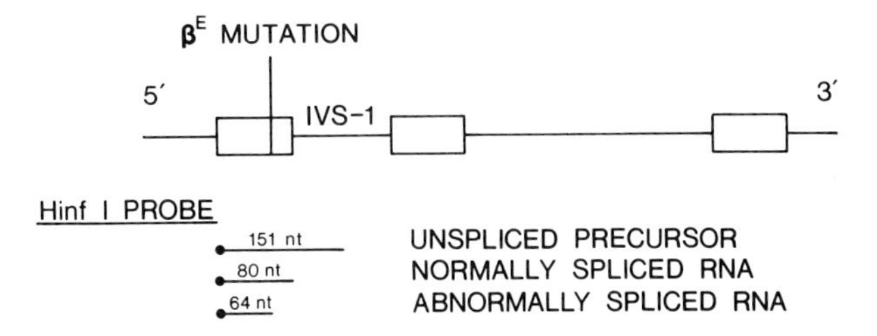

Fig. 3. Expression of the Cloned β^E-Globin Gene in HeLa Cells. Cells were transfected with the normal and β^E-globin genes as described by Treisman *et al.*[5] RNA's were S1 nuclease mapped as detailed in Figure 2. Retic = reticulocyte RNA. Note the abnormally spliced β^E RNA and the increased levels of β^E precursor RNA.

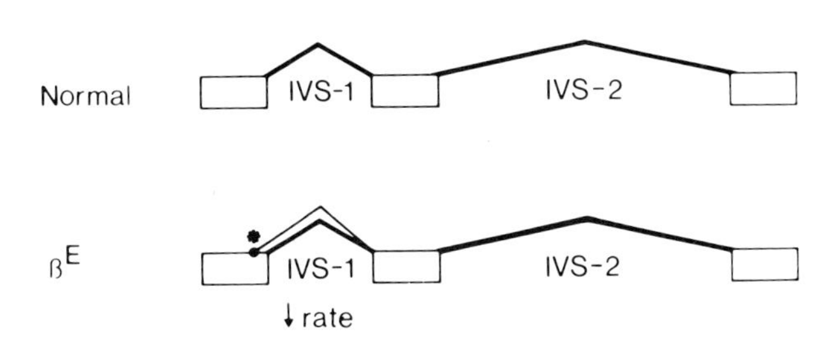

Fig. 4. PATTERNS OF RNA PROCESSING Processing of the β^E transcripts are processed into exon-1 neighboring the β^E mutation (the asterisk) and the excision of IVS-1 is delayed.

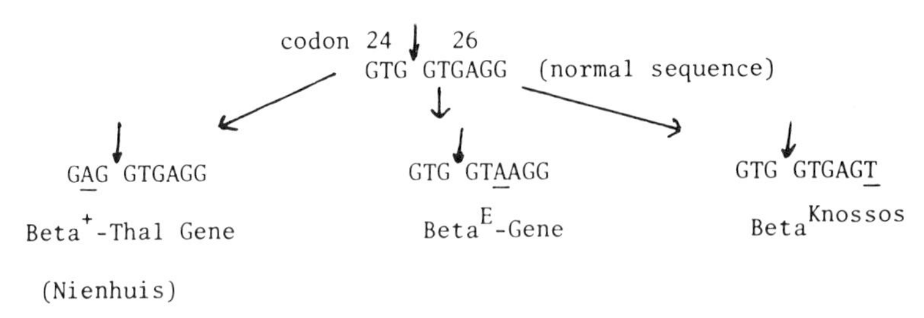

Fig. 5. MUTATIONS OF THE EXON–1 CRYPTIC SITE ASSOCIATED WITH A THALASSEMIC PHENOTYPE The β^+-thalassemic gene to the left is reported by Humphries *et al.*[17]

mutations that generate new donor-like sequences within IVS-2 have been identified.[1,12] In both instances, a considerable fraction of the RNA transcripts are processed abnormally using this internal donor site.

The fourth class of RNA processing mutants was entirely unsuspected prior to study of specific mutant globin genes. In this group are genes with nucleotide substitutions within a *coding* region that alter RNA processing. One example that we have extensively studied is the β-chain variant of HbE. This β-chain mutant has an amino acid replacement of lysine for glutamic acid in position 26 of the β-polypeptide. The β^E-globin chain is underproduced in erythroid cells and its mRNA is deficient relative to α-mRNA, a feature characteristic of β-thalassemias rather than simple hemoglobinopathies.[13,14] To investigate this unusual phenotype we have cloned and expressed β^E-genes. The expected G→A nucleotide substitution in codon 26 was found in two apparently independently arising β^E-genes and no other mutations were observed.[15,16] B^E-globin RNA transcripts in HeLa cells are processed abnormally in two ways:

1. Excision of IVS-1 is delayed, and
2. Some transcripts are processed alternatively into a cryptic-donor site that neighbors the codon 26 mutation[16] (Figure 3).

Therefore, the mutation in the gene enhances the affinity of a preexisting cryptic-donor site for the splicing apparatus. This enhanced donor-site competes, albeit somewhat weakly, with the normal site in splicing (Figure 4). Humphries *et al*[17] have observed a similar and more pronounced effect of a silent nucleotide substitution in codon 24 on splicing at the same position in the gene of a Black β^+-thalassemic (Figure 5). Furthermore, an underproduced β-chain variant, Knosses,[18] with a substitution in codon 27 has been discovered that may also affect RNA processing by activation of the same cryptic-donor site (Figure 5). Mutations in coding sequences that adversely effect normal RNA processing comprise a new class of gene dysfunction that may be significant in the pathology of other inherited diseases.

MUTATIONS OF GENE TRANSCRIPTION

Defects of gene transcription appear to be quite rare in the thalassemias. However, two different mutant genes within this class have been identified. The first, found as a rare allele among Mediterraneans, has a C→G substitution 87 nucleotides upstream from the cap site or start of RNA transcription.[1] This substitution lies 11 base pairs upstream from the CAT box homology[19] that resides within sequences normally required for high level gene transcription as defined by studies with the rabbit β-globin gene and the herpes thymidine kinase gene.[20,21] When studied in the transient expression in HeLa cells, this gene displays an approximate ten-fold reduction in its capacity to direct synthesis of β-mRNA.[7] *In vitro* manipulations with this gene have demonstrated that this abnormal phenotype is due to the −87 substitution rather than another unknown change elsewhere in the DNA. Poncz *et al*[22] have observed a nucleotide substitution within the TATA box homology[19] in a Kurdish β-thalassemic gene. The precise effects of this substitution on gene expression are not as yet characterized. These mutations in β-thalassemia genes with regions previously defined as essential for gene transcription in various test systems are particularly significant as they serve to establish the *in vivo* relevance of these segments to normal gene expression *in vivo* within erythroid cells. The mechanism by which these mutations exert their down-regulation effects is not known.

DIRECT DETECTION OF GENE MUTATIONS IN DNA

Besides providing a detailed molecular description of β-thalassemia, elucidation of the specific mutations in this disorder should contribute to improved prenatal diagnosis of thalassemia. Currently prenatal diagnosis of hemoglobinopathies can be achieved by analysis of globin chain synthesis in fetal blood samples.[23] Although highly useful, the method carries an approximate five percent risk to the fetus due to the procedures required for fetal blood sampling. Diagnosis by DNA analysis would be preferable. Linkage analysis, as originally pioneered by Kan and Dozy,[24] is often useful in detection of β-thalassemia but usually requires extensive and time-consuming family studies.[25] Direct detection of specific mutations would be advantageous.

Knowledge of specific substitutions in β-thalassemia genes now permits use of the detection method developed by Wallace and coworkers in which synthetic DNA fragments (oligonucleotides) are employed in blot hybridization.[26,27] Wallace *et al* have shown that single nucleotide mismatches between a synthetic DNA probe and a cloned gene are detectable by hybridization under carefully controlled conditions. We have employed this approach to develop an assay for the common β^+-thalassemia gene of Mediterraneans in uncloned DNA samples.[28] This particular thalassemia gene accounts for about two-thirds of Greek β-thalassemia genes and a

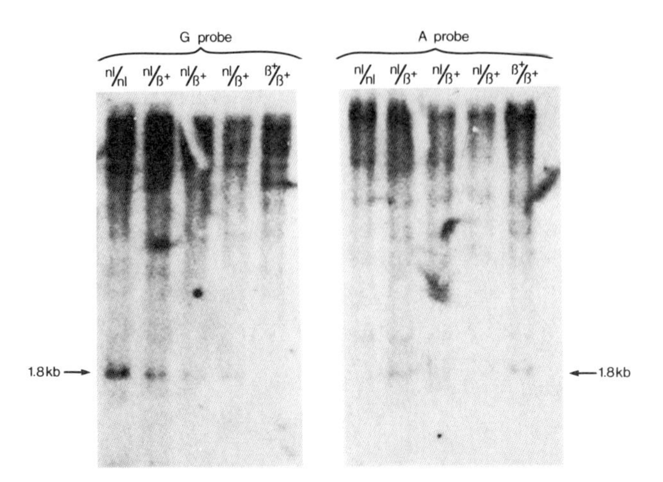

Fig. 6. Detection of the Common β^+-Thalassemia Mutation with Synthetic Oligonucleotides. Cellular DNA's with the genotypes indicated at the top of each lane were digested with *BamHI,* eloctrophoresed in agarose, and then blotted to nitrocellulose.[29] One filter, on the left, was hybridized with a 19-mer corresponding to the normal IVS-1 sequence in the region of the β^+-thalassemia position 110 mutation. A replicate filter, on the right, has hybridized with the mutant synthetic 19-mer. The region of interest is 1.8 kb, shown by the arrows. The synthetic probes only detect their homologous sequences in these samples and permit direct genotype assignments (see ref. 28).

third of Italian genes.[1] Synthetic 19 nucleotide long DNA probes corresponding to the normal and mutant gene sequences spanning the region of IVS-1, that is altered in this particular gene, were prepared. For simplicity the normal oligonucleotide has been termed the G probe and the mutant sequence the A probe, as the mutation at position 110 of IVS-1 is G→A. Conditions were established with cloned normal and β-thalassemic genes in which these probes would hybridize stably only with their homologous sequences in blot hybridizations using the method of Southern.[29] Hybridization of these probes separately to blots containing restriction digests of uncloned DNA's (Figure 6) allows direct assessment of genotypes. With an appropriate battery of synthetic probes all β-thalassemia mutations should be directly dectectable in this manner in amniotic fluid cells DNA's. In the future this approach should appreciably improve prenatal diagnosis of these disorders.

SUMMARY

The β-thalassemias represent the first inherited human disorders in which systematic examination has nearly completely elucidated the underlying molecular basis. These studies have demonstrated the remarkable genetic heterogeneity of this condition and revealed unsuspected insights in the complexity of RNA processing and the *in vivo* significance of presumed promoter elements upstream from the β-globin gene. Finally, characterization of specific molecular defects will ultimately yield practical benefits in improved prenatal diagnosis. This comprehensive approach to molecular analysis of a disease is likely to be a prototype for investigation of other inherited disorders of man.

ACKNOWLEDGMENT

Research in the laboratory was supported by grants from the National Institutes of Health, the March of Dimes and by a Research Career Development Grant of the National Institutes of Health.

Figure 4 reprinted by permission from *Nature*, Vol. 300. 1982; 5894:768-69. Macmillan Journals Ltd.

REFERENCES

1. Orkin SH, *et al.* Linkage of β-thalassemia mutations and β-globin gene polymorphisms with DNA polymorphisms in the human β-globin gene cluster. *Nature* 1982; 296:627-31.
2. Antonarakis SE, Boehm CD, Giardina PJV, Kazazian HH jr. Nonrandom association of polymorphic restriction sites in the β-globin gene cluster. *Proc Natl Acad Sci USA* 1982; 79:137-41.
3. Banerju J, Rusconi S, Schaffner W. Expression of a β-globin gene is enhanced by remote SV40 DNA sequences. *Cell* 1981; 27:299-308.
4. Felber BK, Orkin SH, Hamer DH. Abnormal TNA splicing causes one form of α-thalassemia. *Cell* 1982; 29:895-902.
5. Treisman RA, Proudfoot NJ, Shander M, Maniatis T. A single-base change at a splice site in a β^0-thalassemic gene causes abnormal RNA splicing. *Cell* 1982; 29:903-11.
6. Leder P. Discontinuous genes. *N Engl J Med* 1978; 298:1079-081.
7. Treisman RA, Orkin SH, Maniatis T. Submitted *Nature,* 1983.
8. Spritz RA, *et al.* Base substitution in an intervening sequence of a β^+-thalassemic human globin gene. *Proc Natl Acad Sci USA* 1981; 78:2455-459.
9. Westaway D, Williamson R. An intron nucleotide sequence variant in a cloned β^+-thalassemia gene. *Nucleic Acids Res* 1981; 9:1777-788.
10. Busslinger M, Moschonas N, Flavell RA. β^+-thalassemia: aberrant splicing results from a single point mutation in an intron. *Cell* 1981; 27:289-98.
11. Fukamaki Y, Ghosh PK, Benz EJ Jr, Lebowitz P, Forget BG, Weissman SM. Abnormally spliced messenger RNA in erythroid cells from patients with β^+-thalassemia and monkey kidney cells expressing a cloned β^+-thalassemic gene. *Cell* 1982; 28:585-93.
12. Spence SE, Pergolizzi RG, Donovan-Pelluso M, Kosche KA, Dobkin CS, Bank A. Five nucleotide changes in the large intervening sequence of a β-globin gene in a β^+-thalassemia patient. *Nucleic Acids Res* 1982; 10:1283-290.
13. Benz EJ Jr, *et al* Molecular analysis of the β-thalassemia phenotype associated with the inheritance of HbE ($\alpha_2\beta_2{}^{26\ glu \rightarrow lys}$). *J Clin Invest* 1981; 68:118-26.
14. Traeger J, Wood WG, Clegg JB, Weatherall DJ. Defective synthesis of HbE is due to reduced levels of β^E mRNA. *Nature* 1980; 288:497-99.
15. Antonarakis SE, *et al.* Evidence for multiple origins of the β^E-globin gene in Southeast Asia. *Proc Natl Acad Sci USA* 1982; 79:6608-11.

16. Orkin SH, Kazazian HH Jr, Antonarakis SE, Ostrer H, Goff SC, Sexton JP. Abnormal RNA processing due to the exon mutation of the β^E-globin gene. *Nature* 1982; 300: 768-69.
17. Humphries RK, Ley TJ, Goldsmith ME, Cline A, Kantor JA Nienhuis AW. Silent nucleotide substitution in a β^+-thalassemia gene activates a cryptic splice site in a globin RNA coding sequence. *Blood* 1982; 60:54a.
18. Arous N, *et al.* Hemoglobin Knossos, 27 β Ala→Ser (B9): A new hemoglobinopathy presenting as a silent β-thalassemia. *Blood* 1982; 60:51a.;
19. Efstratiadis A, *et al.* The structure and evolution of the human β-globin gene family. *Cell* 1980; 21:653-88.
20. Grosveld GC, deBoer E, Shewmaker CK, Flavell RA. DNA sequences necessary for transcription of the rabbit β-globin gene *in vivo. Nature* 1982; 295:120-25.
21. McKnight SL, Kingsbury R. Transcriptional control signals of a eukaryotic protein-coding gene. *Science* 1982; 217:316-24.
22. Poncz M, Ballantine M, Solowiejczk D, Barak I, Schwartz E, Surrey S. β-thalassemia in a Kurdish Jew. *J Biol Chem* 1982; 257:5995-996.
23. Alter BP. Prenatal diagnosis for haemoglobinopathies: a status report. *Lancet* 1978; 2: 1152-155.
24. Kan YW, Dozy AM. Polymorphism of DNA sequence adjacent to the human β-globin structural gene: relation to sickle mutation. *Proc Natl Acad Sci USA* 1978; 75:5631-635.
25. Kazazian HH Jr, Phillips JA III, Boehm CD, Vic TA, Mahoney MJ, Ritchey AK. Prenatal diagnosis of β-thalassemias by amniocentesis: linkage analysis using multiple polymorphic restriction endonuclease sites. *Blood* 1980; 56:926-30.
26. Wallace RB, Schold M, Johnson MJ, Dembek P, Itakma K. Oligonucleotide directed mutagenesis of the human β-globin gene: a general method for producing specific point mutations in cloned DNA. *Nucleic Acids Res* 1981; 9:3647-656.
27. Conner B, Wallace RB, Teplitz RL. Analysis of gene dosage of sickle cell β-globin (β^S) in human genomic DNA by DNA hybridization with synthetic oligonucleotides. *Clin Res* 1982; 30:46A.
28. Orkin SH, Markham AF, Kazazian HH Jr. Direct detection of the common Mediterranean β-thalassemia gene with synthetic DNA probes: An alternative approach for prenatal diagnosis. *J Clin Invest* 1983; in press.
29. Southern EM. Detection of specific sequences among DNA fragments separated by gel electrophoresis. *J Mol Biol* 1975; 98:503-17.
30. Chang JC, Kan YW, β^0-thalassemia, a nonsense mutation in man. *Proc Natl Acad Sci USA* 1979; 76:2886-889.
31. Orkin SH, Goff SC. Nonsense and frameshift mutations in β^0-thalassemia detected in cloned β-globin genes. *J Biol Chem* 1981; 256:9782-784.
32. Moschonas N, deBoer E, Grosveld FG, Dahl HH, Shewmaker CK, Flavell RA. Structure and expression of a cloned β^0-thalassemic globin gene. *Nucleic Acids Res* 1981; 9:4391-401.
33. Trecartin RF, Liebhaber SA, Chang JC, Lee KY, Kan YW. β^0-thalassemia in Sardinia caused by a nonsense mutation. *J Clin Invest* 1981; 68:1012-017.
34. Kinniburgh AJ, Maquat LE, Schedl T, Rachmilewitz E, Ross J. mRNA-deficient β^0-thalassemia results from a single nucleotide deletion. *Nucleic Acids Res* 1982; 10:5421-27.
35. Breathnach R, Benoist C, O'Hara K, Gannon F, Chambon P. Ovalbumin gene: Evidence for a leader sequence in mRNA and DNA sequences at the exon-intron boundaries. *Proc Natl Acad Sci USA* 1978; 75:4853-857.

INFERRING VARIATIONS IN GENOTYPE FROM PROPERTIES OF DNA

Leonard S. Lerman
Stuart G. Fischer
Nadya Lumelsky

Our current revolution in genetics has seen the replacement of the classical genome cast in rigid typemetal punching away moronically throughout somatic development with an amazingly supple, versatile genome undergoing shifting, revision, and gross, wholesale editing. The notion of a universal, uniform protein repertory, together with a hypothetical genetic load consisting of a few lethal recessives per haploid genome, together with a larger number of more benign differences has been replaced by appreciation of a high level of diversity and individuality. In the classical definition, a difference in genotype owed its existence conceptually to the analysis of variations in phenotype. Now, the genome is dense with sites of variability for which the phenotypic counterparts, if any, are unknown.

There are diverse ways in which genotype has become more familiar than phenotype. For example, where the phenotype is recognized as the sequence of amino acid residues in a protein, it is sometimes better to deduce the amino acid sequence from the base sequence in DNA, rather than to try to purify and analyze a rare or unobtainable protein.

In contemplating the variability of the human genome and its consequences, it is reasonable to ask: How can we deal with the analysis of a genome of three billion nucleotides? Restriction endonucleases generate 10^6 to 10^7 fragments. The mRNA's of many of the proteins are so rare or transient as to defeat isolation. Sequencing the wild type alone can be estimated to require three million doctoral dissertations. To focus on a specific problem in human genetics: Suppose that an alteration in genotype is roughly localized, and that the DNA fragments of both the wild and the aberrant type are available for study. One approach to finding the aberration has been described here. (White, 1983) If a base substitution or an insertion or a deletion occurs at the point in the sequence that serves as the site for cleavage by a restriction endonuclease, cleavage that would be expected with the wild type sequence will not take place in the mutant, and two fragments will

ISBN 0-12-492220-1

be produced with different lengths and different boundaries. Some changes will introduce new restriction endonuclease sites, and a fragment that would otherwise have been a product of cleavage disappears, giving two new, shorter fragments. Occasionally, the specificity of one or another restriction endonuclease is appropriate for detecting an interesting and prevalent alteration of sequence. For example, the endonucleases *DdeI* and *MstII* normally cleave the β-globin sequence. The substitution responsible for sickle cell anemia changes an A into a T at base 17 (from the 5′ end of the gene), changing glutamate to valine and therefore no longer matching the specificity of either enzyme. The result is a distinctly different and diagnostic fragmentation pattern.

However, there are serious limitations. Sixteen separate tests with each of the full set of 16 different restriction nucleases that have the usual pattern of specificity for four adjacent bases would provide only a six percent chance of detecting a random base change. More of the genome could be scrutinized with other enzymes, but the number of enzymes and separate tests required would need to be many fold larger than the number available today if a reasonable level of certainty in detecting the changes were to be assured. We may also note that the presence or absence of fragmentation does not specify what kind of change is present. It does not even guarantee that the change is confined to one or a very few nucleotides.

During the last several years, our laboratory has been engaged in a study of the relation between base sequence, melting of DNA, and the ability of partly-melted DNA to be transported by an electric field in an aqueous gel. We find that the mobility is so strongly dependent on structure, and that this is in turn determined by sequence, that we have a new and exceedingly sensitive means to explore the physical properties and stability of DNA on the one hand, and to detect changes in sequence on the other. At the same time, it has also become possible to bring about a clear physical separation of DNA molecules nondestructively according to small or large changes in sequence. The observations are related to sequence in a way that permits substantial inference as to the nature of the changes. We think that it will provide a particularly useful tool for the detection of genetic aberrations in humans. Because it makes possible the examination of relatively large samples of sequences in relatively simple laboratory operations, it can permit the localization of unknown changes within well-defined small sections. The single small section carrying the change can be isolated and subjected to conventional sequence analysis. The possibility of physical separation can make cloning of the altered sequence feasible and simple.

Because of the novelty of the system and our growing capability to establish theoretical correspondence between sequence and the experimental observations, it will be useful to examine in some detail the system and the principles on which it is based.

If DNA is heated under close temperature control, it is found that there is an equilibrium at each temperature in which part of the molecule retains orderly, double-helical characteristics while other parts of the molecule have lost their helicity. The complementary strands of DNA in the non-helical patterns become unravelled and adopt a randomized, irregular configuration. At ordinary physiological temperatures in aqueous solution with salts, the equilibrium strongly favors the double helical form, while at elevated temperatures (for example, near boiling) the equilibrium strongly favors unravelling and disorder. This change is given the short-hand term "melting" in the sense that melting commonly implies a change from the orderly arrangement of molecules in a crystal to a more random but still condensed collection in a liquid. However, the term does *not* imply that separate bases in one strand of the double helix lose the physical connections that maintain continuity of the strand.

Melting of a crystal does not progress gradually over a long interval of temperatures, but, if the substance is pure, the change from crystal to liquid takes place at an indefinitely narrow temperature range. Thus, water ever-so-slightly below 0°C is all liquid. The melting of DNA is similar, but not exactly the same, in that DNA is orderly in only one dimension, while crystals like ice or any ordinary substance are orderly in three dimensions.

In 1945, long before any serious structural studies of DNA had been undertaken, E. Schroedinger[1] suggested in his influential book, *What Is Life*, that DNA be regarded as a one dimensional crystal. We must realize also that the units forming the crystal-like pattern of DNA are not identical, as they would be in ice, sugar, steel, *etc.* Since there are four different types in scrambled order, the double-helix does not melt as sharply as a crystal, but each local region over a length of perhaps 50 to 500 bases changes from helix to disorder over an extremely narrow temperature span. Adjacent regions may undergo their transitions at quite different temperatures, and the melting of the entire molecule proceeds as a series of jumps. First, one region will melt with all of the bases in that group of neighbors melting together. A small further increase in temperature may have no effect until, at a higher level another group will melt, and so on. This progression in groups rather than individually is called the cooperativity of melting.

The joint melting of a large part of the sequence within a very narrow temperature span is all the more striking when one considers the broad diversity that would be expected in the melting temperatures of clusters of 2, 3, or 4 base pairs, if their properties were observed in isolation and not embedded in a helix. The parameters that enter our calculations are specified in terms of adjacent base pairs in which the order of the bases must be considered. There are ten different ways in which neighboring base pairs can be composed. The number is only ten because some neighboring pairs are equivalent. For example, the doublet described by G following A on one strand is identical with the doublet in which T follows C. An AT next to a TA is substantially less stable than AT adjacent to another AT. Each of the ten doublets of the ten has its own local characteristic melting temperature, with a span of 100°C between the highest and lowest. These values jump up and down along the sequence in natural DNA. How, then, does the cooperativity exert such a dramatic leveling? As a reasonably apt analogy, we can consider a line of persons of varying strength and stature confronting a severe disturbance, such as an earthquake or a strong wind. We see such a line sketched at the top of Figure 1. If the disturbing influence – in our experiment, temperature, here the trembling of the earth or the wind indicated by the wavy line – increases a little, the least sturdy fall. As the strength of the disturbance increases, and if each individual reacts independently as indicated on the left side of the sketch, there would be a gradual progression of collapse, one by one, until with the strongest disturbance none remains standing. However, suppose the same individuals had locked their arms so each individual in the line may stand or fall depending on the status of his neighbors as sketched on the right. If there is a slight disturbance, the sturdy members can support the feebler everywhere along the line except, perhaps, at the ends because end members have only one arm linked. The sketch suggests that one end of the line has a better selection of sturdy members. As the intensity of the disturbance increases, the weaker end of the line collapses as a group. Because of the cooperativity, it is unlikely that any individual can fall down alone, but if too many are incapable of standing, the entire group falls. The other end of the line can remain standing as a group until the conditions become significantly more severe. This implies also that the united group will fall under circumstances that the strongest alone might have been able to withstand.

A reasonably good understanding of the melting process has been developed from the standpoint of statistical mechanics, although we must note that the theory has not yet been subjected to extensive comparison with experiments. These studies will provide some of that comparison.

The theoretical understanding is built on a few simple assumptions. The only interactions in the double helix that need to be considered are those between each base pair and its nearest neighbors. Each doublet of two adjacent base pairs can be thought to have an intrinsic stability, a number that would describe their tendency to remain paired and stacked in contact with each other in the helix. Each position along the helix can be described by an appropriate value within the set of ten possible types of doublets. The chance that one base pair would become nonhelical, that is, neither lying flat on top of the adjacent base pair nor retaining its internal hydrogen bonding, but hydrogen-bonded to water instead, is influenced by whether its neighbor is stacked and paired. The strength of the nearest neighbor correlation is described by a single number called the cooperativity parameter. In addition, it is necessary to take into account a geometrical effect – those bases in disordered regions that are bounded by helix at both ends are more likely to encounter each other and, therefore, to form helices than bases in disordered regions at the end of the molecule bounded by helix in only one direction. The theoretical formulations of Poland,[2] Fixman and Friere[3] together with estimates of the values of the relevant numerical parameters provided by Gotoh and Tagashira[4] and

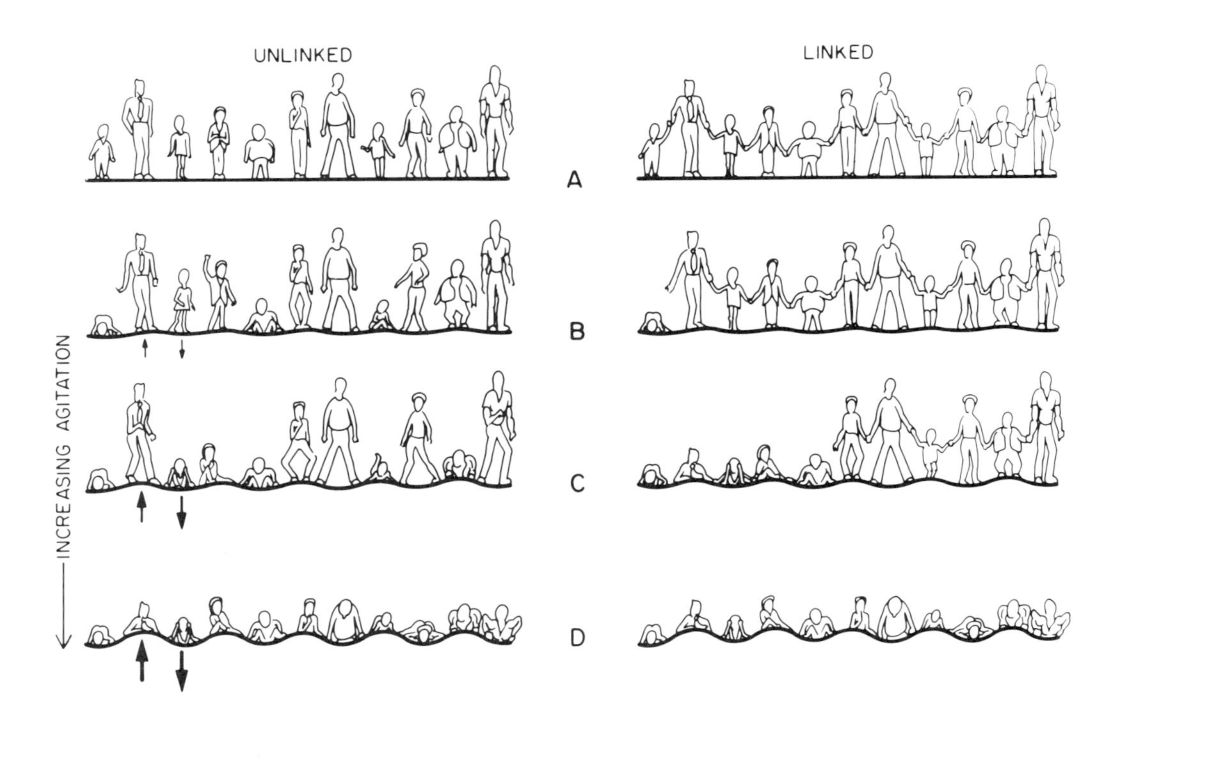

Fig. 1. Cooperativity as a consequence of strong nearest neighbor interactions. From top to bottom: a linear array of elements of different individual stabilities is subjected to increasing intensity of agitation. On the left, the behavior of each element in the array is independent of any other; on the right, the response to agitation is determined partly by the response of the immediately neighboring elements.

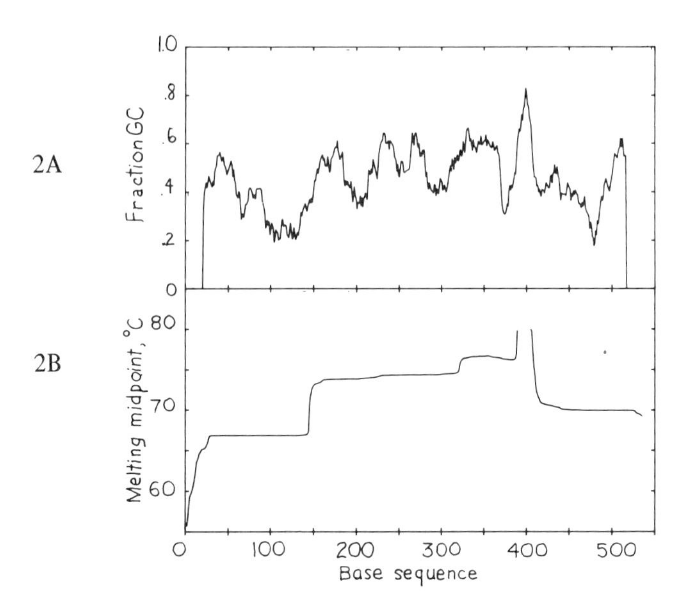

Fig. 2. Upper: The distribution of AT and GC base pairs along a fragment of λ DNA. The fragment consists of the sequence between bases 39125 and 39560 (inclusive), according to conventional numbering. Each base pair is assigned a binary value, 1 or 0, for GC or AT respectively. The local density of GC pairs is presented as a smoothed function using a sliding cubic polynomial fit to each set of 21 contiguous base pairs. The base sequence is taken from Wulff and Rosenberg.[7] Lower: Melting map of the λ fragments. The nearest neighbor stability values for the base sequence of the *lambda* fragment have been introduced into the statistical-mechanical theory of cooperative melting formulated by Poland (1974) and Fixman and Friere (1977), using conventionally acceptable values of the cooperativity and loop entropy parameters. The plot shows the contour of the 50% temperature for local melting equilibrium, the temperature at which each base pair along the sequence is equally likely to be helical or disordered. The change to nearly complete helicity or nearly complete melting occurs within less than one degree C on either side of the 50% curve. Comparison with the upper plot shows that regions of higher GC density undergo the melting transition at higher temperatures, but the cooperativity and the tendency to suppress internally disordered regions result in groupwise melting in domains, despite large, short-range compositional variations. The stability values for the different adjacent base pair doublets were taken from Gotoh and Tagashira.[4]

others, have made it possible (with a non-trivial reliance on a digital computer) to make a direct, theoretical calculation of the progression of melting as the temperature is raised for any given sequence of bases in a DNA molecule. The calculation provides a formidable amount of still-somewhat-hypothetical information about the progression. To select a manageable, descriptive sample, we plot the temperature at which each individual base pair is in precisely the middle of its equilibrium between helix and disorder for the entire sequence. We term this diagram a melting map. A representative pattern is shown in Figure 2B, which was calculated for a 536 base pair DNA molecule that constitutes part of the chromosome of the bacteriophage *lambda*. Where the plotted line lies below a specified temperature, that part of the molecule will be helical at that temperature. The map shows that as temperature rises, melting begins at the left end and gradually proceeds to the right. In the vicinity of 67.8°, there is an abrupt melting of all the bases from 40 to about 145, starting just above 66° and essentially completed below 68°. At 71°, there is abrupt melting from the right hand end to base 470. Complications related to the possibility of complete separation of the strands and the need to invoke bimolecular equilibria may set in at the higher temperatures. The graph at the top, 2A, represents the local variation in base composition along the molecule. The fractional composition in terms of A+T or G+C characterizes the sequence in a simple way and approximates the local stability that would be given by the doublet parameters if cooperativity and loop effect were neglected. This plot can be taken to represent the noncooperative process in Figure 1A, while the melting map below corresponds to Figure 1B. Comparison of 2A and 2B shows how difficult it is to guess the melting behavior from the base sequence without benefit of theory.

Gel electrophoresis has provided the means by which the length of DNA molecules can be determined with modest accuracy and constitutes an essential part of the restriction endonuclease procedure for the detection of base substitutions. The gels in common use can be regarded as three-dimensional networks with water-filled pores larger than the diameter of the double helix but much smaller than either the length of the molecule or the statistical diameter of an average tangle of a long, semi-flexible thread-like molecule. Thus, the DNA molecule must be entwined through many pores as a gently curving random chain. The electric field in electrophoresis is far too weak to bring about a significant orientation of the molecule in the direction that it is supposed to move. Accurate measurements of the relationship between a molecule's length and its electrophoretic mobility, together with a straightforward theoretical explanation based on molecular geometry of why long molecules move more slowly than short ones, has been presented recently by Hervet and Bean.* The velocity shows very little, if any, sensitivity to the base sequence. In aqueous salt solutions in the absence of a gel, all DNA molecules travel at essentially the same velocity in the electric field, independent of length. What determines mobility in gels becomes complex at the onset of melting,

*Hervet H, Bean C. personal communication

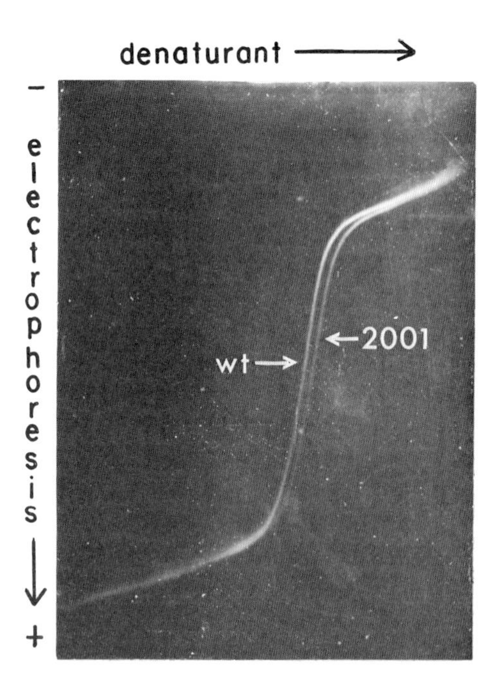

Fig. 3. Electrophoretic mobility of the wild type fragment and the *c* II2001 mutant in a polyacrylamide gel at 60°C as a function of denaturant concentration. A mixture of both fragments began migration in a straight line along the top of the gel. The gel contained a linear gradient of denaturant increasing from 1.4 M urea and 8% (v/v formamide on the left to 3.5 M urea and 20% (v/v) formamide on the right. Top to bottom concentration is nominally uniform at each gradient concentration, but there is a slight loss everywhere along the top. About ½ as much *c* II2001 DNA as wild type was applied.[6]

for example, when bases from the left end to 145 are melted and the remainder is still helical. There is now no longer simply a single random path representing the contour of the molecule in the three-dimensional network; there are additional divergent random paths, corresponding to the contours of the separated strands where they are melted and now independent of each other. The effect of branching is to impose a need for much extra time for the molecule to find its way through the gel. Each branch must be disentangled from its current disposition before the molecule as a whole can advance. This effect is shown strikingly in Figure 3 in a somewhat intricate experiment. DNA has moved through the gel from top to bottom, starting as a uniform sample applied everywhere along the top. However, from left to right, the gel contains a linearly increasing mixture of organic solvents, urea and formamide, which lowers the melting temperature of the double helix in a simple way. Although the experiment is carried out at a constant temperature of 60°, the left to right gradient of urea and formamide is equivalent to a left to right increase of effective temperature in the gel from 60° to about 95°. Any molecule of DNA moving down from the top sees only one effective temperature, but the neighboring molecule to the right sees a slightly higher, uniform effective temperature. The DNA sample in this experiment is the *lambda* fragment whose sequence provided the melting map in Figure 2, together with a smaller amount of an almost identical fragment in which a TA pair of the original is replaced by GC. On the left side of the gel, the DNA has moved almost to the bottom, but, as the urea-formamide concentration increases toward the right, movement is somewhat slower, corresponding to the reduction in velocity due to gradual melting in from the end of the molecule. At about the middle of the gel, there is an abrupt change in its mobility, and movement in the higher denaturant concentrations to the right is very much less.

We can now ask whether the sharp drop in mobility at a particular concentration of denaturant can be identified with the melting of all or part of the molecule. We have obtained a clear answer by the same method that has proved so widely useful in fundamental biology, that of examining the consequences of sequence alteration among a series of closely-spaced mutations in the locus of interest.

However surprising it may seem, the substitution of a single base pair by any of the three other distinct pairs introduces sufficient perturbation in the melting properties to permit detection by the effect on its electrophoretic mobility in a denaturing gel. For the detection of substitutions and most of the other experiments we shall describe, a gradient of denaturing solvent parallel to the direction of electrophoretic migration (rather than perpendicular, as in Figure 3) is useful. The molecule travels into the equivalent of a gradually increasing temperature. When the molecule reaches a characteristic depth in the gradient where melting occurs, its electrophoretic mobility falls, and further movement towards higher equivalent temperatures is either very small or altogether negligible. We can call the depth the retardation point. The electric field can be applied until every molecule reaches its retardation point in the gel. Each establishes a position in a pattern that is relatively insensitive to time or to continued application of the electric field. We have shown[5] that if a variety of molecules of widely varying length carry the same most easily-melted sequence, they are all retarded at very nearly the same depth in the gradient.

Recently we have explored the consequences and site-specificity of single base substitutions[6] using a set of *lambda* strains carrying mutations in the *cy* region, provided by D. Wulff of our university. The base sequence of the region and the specific nature of each substitution (described in Table 1) has been determined by Wulff and M. Rosenberg.[7] The depth in the gradient reached by a fragment of wild-type *lambda* and the corresponding fragments of each of a series of mutants

TABLE 1. Sites of base substitution as *y* region mutants in the *lambda* DNA fragment containing bases 39125 to 39660, inclusive. The sites are numbered from the end of the fragments. Adjacent bases are specified.

Mutant Strain	Sites	Substitution
*cin*1, *cnc*1	84, 88	TGG to TAG
		GTA to GCA
*cy*2001	136	CTT to CGT
*cy*3071	136	CTT to CCT
*cy*3019	139	ACA to ATA
*cir*5-*c*II3086	141, 142	ATAT to ACGT
*c*II3105	144	TGG to TAG
*c*II3059	146	GTT to GAT
*c*II3059-*ctr*2	146, 150	GTT to GAT
		GTG to GCT
*c*II3059-*ctr*1	146, 151	GTT to GAT
		TGC to TAC
*cy*42	153	CAA to CTA
*c*II3114	157	CAA to CGA
*c*II3098	164	AAC to AGC
*c*II3085	167	GAG to GGG
*c*II3638	424	GAT to GGT
*c*II3520	433	CTG to CCG

is shown in Figure 4 as a composite of a number of gels. It is clear that the fragments of many of the mutant phage reach different retardation positions from that of wild type, while some reach levels that are indistinguishable from wild type. The results can be summarized as follows: every base substitution tested that lies between bases 83 and 145 changes the gel position away from wild type, either up or down. Substitutions at positions beyond base 152 fail to shift the gel position away from that of wild type. Although the interpretation of *c*II 3059 and its doubly mutant derivatives *c*II3059-*ctr*1 and *c*II3059-*ctr*2 is not unambiguous (see ref. 6), the data permit identifying this set with other mutations that fail to shift the gel zones. The boundary in the base sequence between mutants that shift and those that fail to shift gel positions then lies between sites 144 (*c*II3071) and 146 (*c*II3059). The correspondence of these results with melting theory can be seen by reference to the maps in Figure 5, which present the difference between the calculated melting temperature of each base in the mutant strain and in wild type. The melting map shows that all of the mutants that change gel positions are substitutions in the lowest melting region at the left end of the fragment. Mutations at positions higher than 144 are in regions calculated to have substantially higher melting temperatures. The steep rise in calculated Tm occurs almost entirely bewteen 144 and 146. Thus, the observations support the supposition that retardation of this fragment in the denaturing gradient is determined solely by its lowest melting region at the left end. This is the first experimental localization of a domain

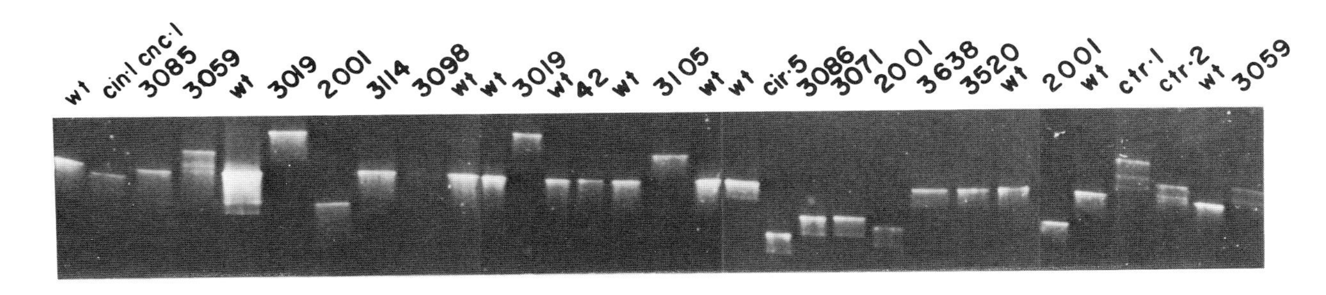

Fig. 4. Comparison of depths in the denaturing gradient of the specified wild type λ DNA fragment and corresponding fragments of the same length from mutant strains carrying single or double base substitutions. Photographs of the fluorescence of four ethidium-stained gels were aligned at the level of the wild type bands.

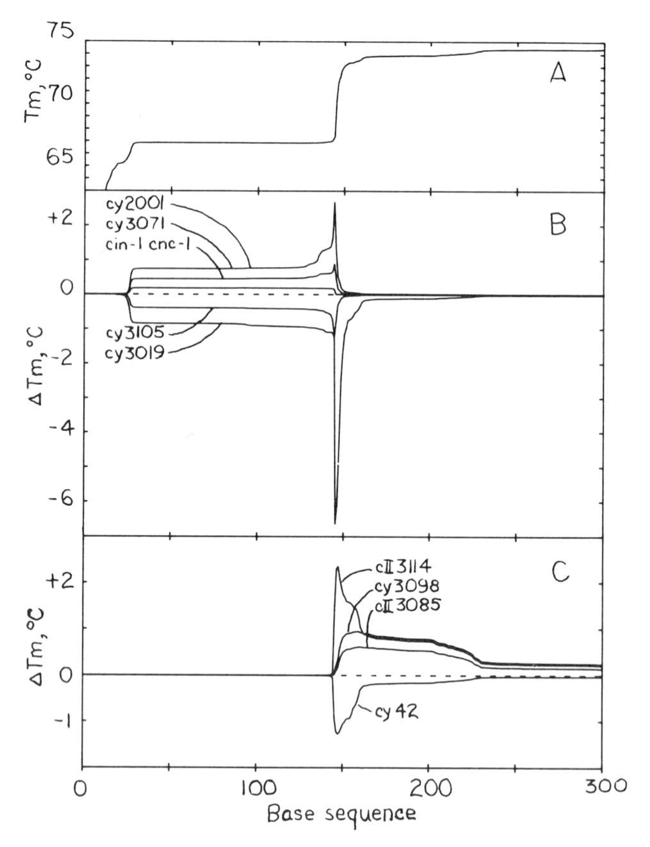

Fig. 5. A. The left end of the map in 2B is reproduced. B & C. Melting difference maps. Complete melting maps for each of the mutant fragments were calculated in the same way, and the map of wild type was subtracted. The plot shows the position and temperature perturbation for melting of the part of each sequence affected by the base substitution.[6]

---

boundary and a demonstration of its agreement with theory. We may note that the calculation has drawn on statistical mechanical theory together with a few measurements made with simple polymers and an *E. coli* plasmid; there is no dependence in any way on information derived from this DNA fragment except its base sequence. If the fragment is further shortened, removing the left end by a cut between positions 235-6, the two substitutions *c*II3638 and *c*II3520 at positions 424 and 433 that are inconsequential in the original, intact fragment alter the retardation level of the truncated fragment in the same way. Figure 6 presents a comparison between the denaturing gradient results and statistical mechanical theory.

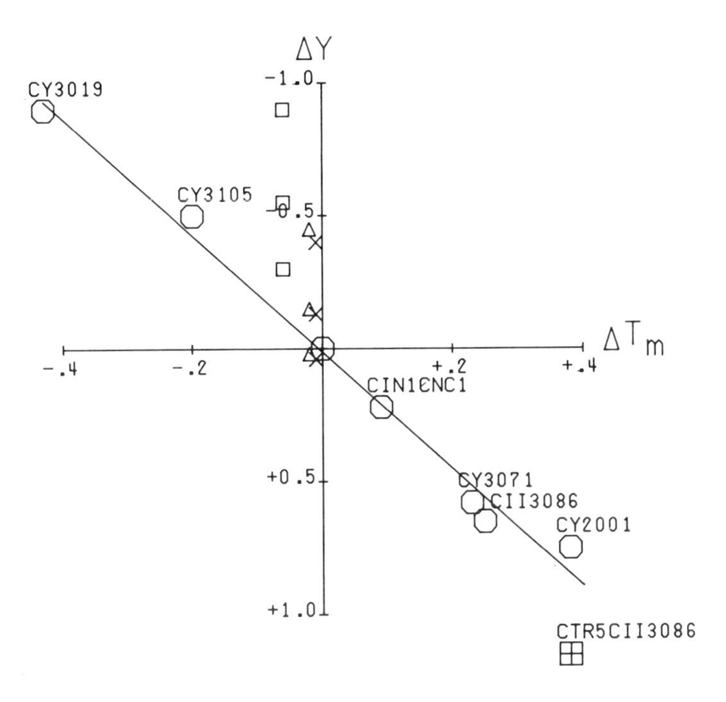

Fig. 6. Correlation of the first domain differences in the melting maps with the experimental gradient positions of mutant fragments. The difference in gradient positions between each fragment and wild type is plotted as the ordinate with the calculated temperature elevation or depression of the first domain as the abcissa. The point at the origin represents the set of substitutions above 148 that are retarded at the level of wild type. The double mutant *ctr*5-*c*II3086 is not included in calculation of the regression line because of more severe effects on the length of the domain.[6]

According to the calculation, each base substitution of this set can be expected to raise or lower the level of the plateau at the left end of the melting map by a small amount. The cooperativity of the interaction between bases spreads the effect of a single substitution over the whole group of 110 base pairs. The close correlation between the calculated change and gel displacement is shown in the figure. It can be seen that a transversion – a mutation that interchanges the members of a base pair between strands without any change in composition – is detected by both the gels and the calculation. The effect of the same substitution at different sites is also seen to vary according to the context.

These results imply that, in general, a base substitution will be detectable if it lies in the lowest melting domain. The direction and magnitude of the gel displacement provide partial specification of the nature of the substitution, not merely a signal that substitution has occurred.

The gel retardation characteristics determined by the domains of highest melting priority are of interest because they provide close characterization of genomic fragments and provide the basis for a new scheme of chromosome mapping. Since the most easily recognizable features are determined by domains, rather than individual bases, we can regard this aspect as coarse-grained identification and mapping. Some of the markers defined by melting characteristics may be identifiable with sequences for particular biological functions. The experiments offer a strong challenge to melting theory: Can we explain the absolute level of the gradient that effects retardation for a given molecule as well as the small deviations in level treated by the substitution study? If so, can we use observed retardation as a partial characterization of the determinant sequence?

I would like to consider mapping first in a purely qualitative sense. A low melting domain represents a molecular marker that is easily recognized in the gradient gel independent (or almost so) of the length or constitution of the remainder of the molecule. If a molecule is cleaved into fragments that are not too small, one of the fragments will retain the domain of highest melting priority of the original molecule and undergo retardation at the same level as the original in the gradient . In the other fragments, domains that held lower priority in the original assume first priority, permitting the recognition of additional markers. Fragmentation by random processes, such as shear or nonspecific endonucleases, are just as applicable as cleavage at specific restriction sites and, in some ways, more useful. Since the appearance of fragments at levels defined by lower priority domains depends on a cleavage that cuts away any domains of higher priority in the original sequence, random fragmentation can generate a mapping metric according to the relation between the probability of cleavage and the length of sequence between ranking domains.

We have presented a relatively simple application of this principle in shear fragmentation of the 4639 bp linear molecule prepared by opening the circular DNA molecule of the plasmid, pBR322, with a single *Eco*RI cut.[8] Minimal shearing with a rotating blade homogenizer produced a diverse family of fragments that gave a complex pattern in a two-dimensional gel, shown in Figure 7. Here the fragments were first separated according to length by conventional gel electrophoresis in the x direction, then driven down into a gradient of increasing denaturant concentration in the y direction, similar to the gel in Figure 4. Because the same highest priority determinant domain controls all lengths that lack a higher priority domain, the pattern consists of an array of streaks extended in the direction of length heterogeneity, corresponding to each family of shear fragments sharing a common determinant.

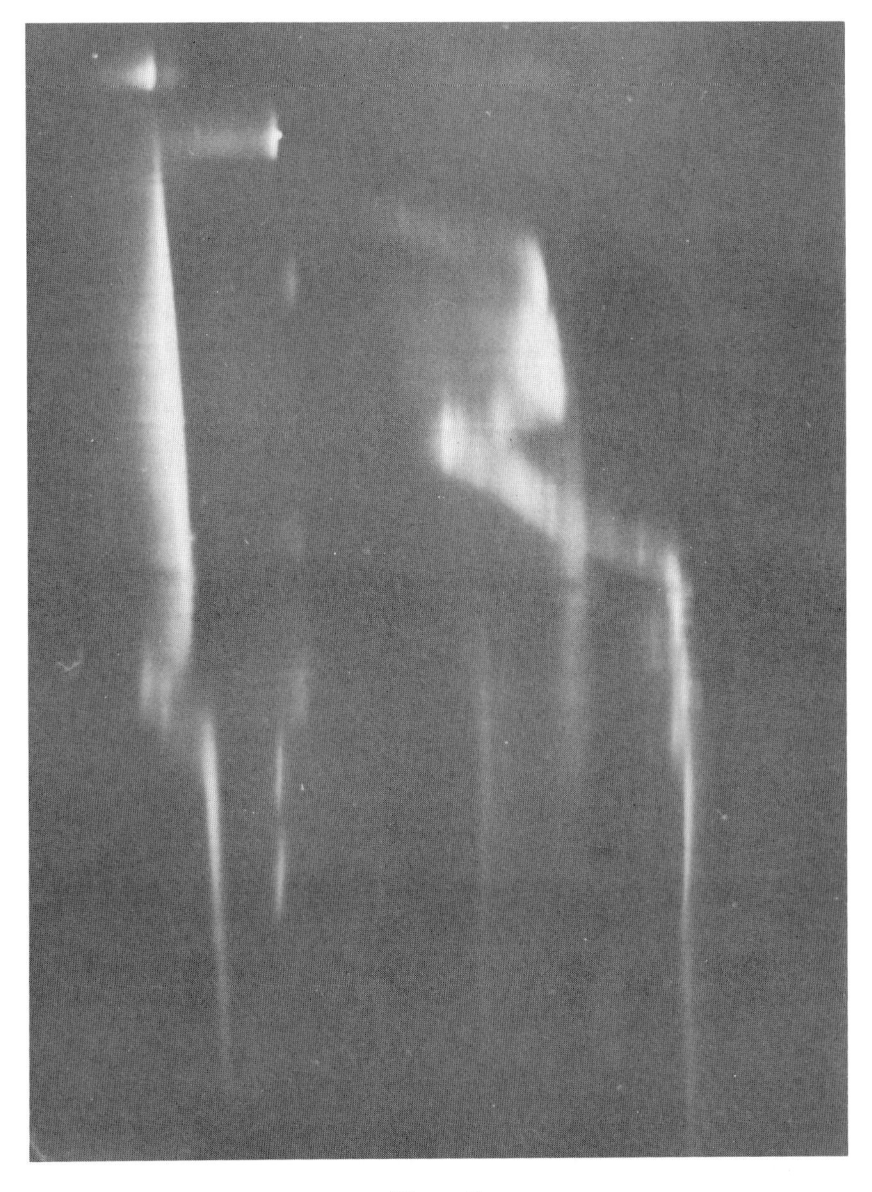

Figure 7

Since there are only two major components distinguished by the gradient at any fragment length, it is clear that nearly all of the original linear molecules have been cut only once by the homogenizer. Here, the rough mapping is simple. Calibration of the system with a substantial set of restriction fragments of the plasmid DNA together with sequence comparisons shows that the upper tier of streaks represents all the fragments from the high numbered end of the molecule and that the lower tier comes entirely from the left end.[9] The pattern displays many features of the variation of sequence along the molecule and permits ordering and localization of certain details. The sites of abrupt increase in local AT density can be specified within the length error of the elctrophoretic x separation. The distinction between interior melting bounded by helical regions and melting inward from the ends, important in melting theory, is also strikingly displayed in the two-dimensional pattern. Because of the clear distinction between the right and left ends of the sheared molecule in the gel pattern, it is possible to construct a detailed map of the probability of cleavage along the sequence by quantitative interpretation of the pattern. In view of the pattern specificity of fragments with one random cut, it may seem less surprising that fully random fragmentation generates intricate but nonetheless well-defined patterns in the two dimensional system.

In approaching sequence characterization of human DNA, where substantial variation beyond single base changes can be expected, it becomes essential to consider how well gradient characterization reflects the relevant sequence. We have applied the denaturing gradient technique to the examination of restriction fragments of human mitochondrial DNA. Mitochondrial DNA is useful as a molecule of manageable size that can be prepared from human sources with adequate purity, and the full sequence of at least one sample has been published. Since it is not too complex, individual fragments are resolved in the gel without the need for transfers and specific probes. It is clear at the outset that we can expect somewhat greater difficulties than with prokaryotic DNA because the limits of fluctuations in composition along the sequence are narrower.

Figure 8 shows the melting maps of four fragments calculated from the base sequence published by the Sanger group.[10] Each fragment can be seen to have a different and characteristic melting progression. Fragment 1 (at the top) is unusual in that the calculation indicates that melting would begin in the interior of the molecule rather than at an end. We are engaged in the first steps in the attempt to understand the details of electrophoresis well enough to arrive at an accurate, purely theoretical calculation of the final gel position of any fragment on the basis of its sequence. With certain reasonable assumptions based on polymer theory, we can use the calculated melting progression to arrive at an expected travel schedule, showing the calculated position of a fragment in the gradient gel as a function of time. Plots of calculations of this sort are shown for the four mitochondrial fragments presented in Figure 9. The plots show that each fragment starts moving down the gel at nearly constant velocity with its rate determined principally by its length.

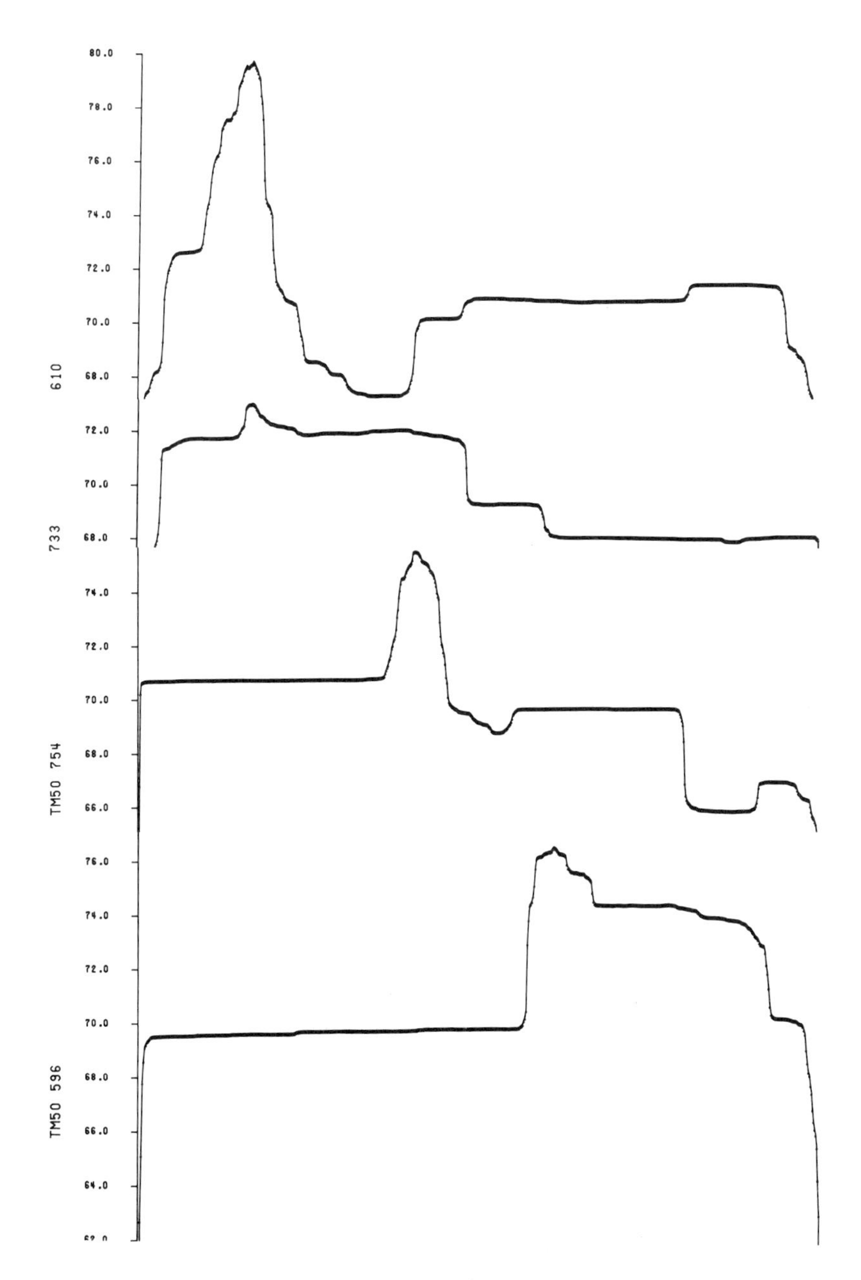

Fig. 8. Melting maps of four fragments of human mitochondrial DNA. The maps were calculated from the sequences determined by Anderson *et al*[10] for the DNA of human placental mitochondria. The actual lengths are 754, 733, 610 and 596, top to bottom; they have been normalized to 100% for this plot.

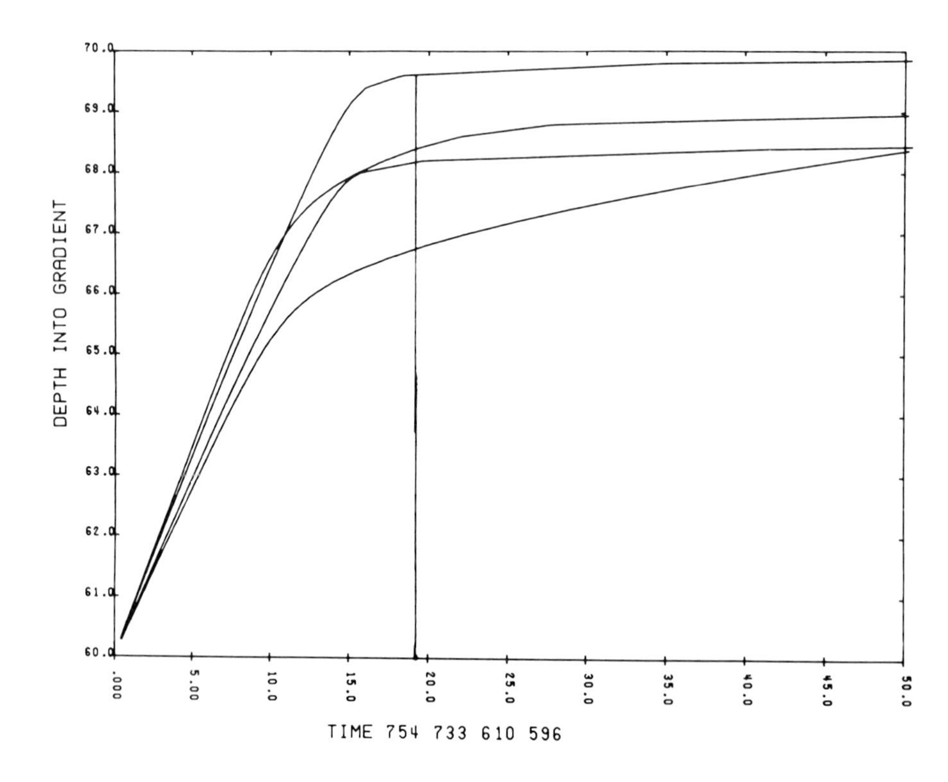

Fig. 9. Calculated movement of the mitochondrial DNA fragments during electrophoresis into an increasing urea-formamide gradient. The change in velocity as the first domain of each fragment melts was calculated, as suggested,[11] as an exponential decline in mobility with melted length using 60 bases as the relevant statistical segment length.

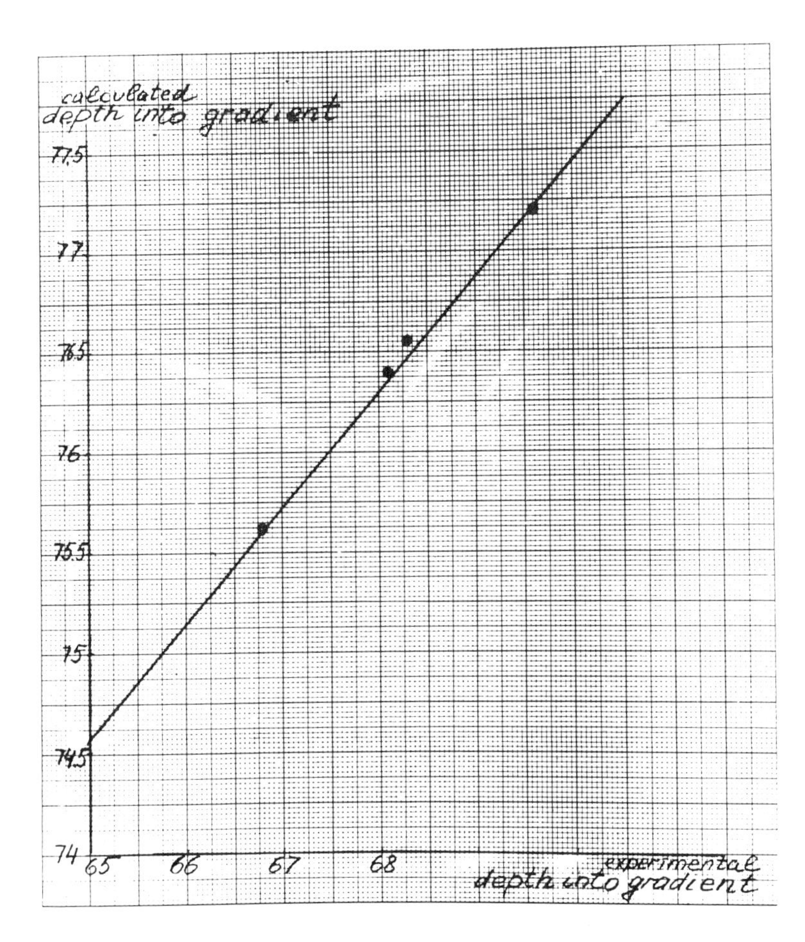

Fig. 10. Comparison of the calculated and experimental depth into the gel for the different restriction enzyme fragments. Correlation of the calculated positions with experimental gradient depths of HeLa mitochondrial DNA fragments after 17 hours in the gradient.

However, at characteristic depths, the velocity of each fragment drops as expected from perpendicular gradient patterns, as we have shown directly with *lambda* fragments.[9] If a relatively long section has melted cooperatively, the fragment is abruptly retarded to a relatively low velocity. Other molecules may decline more gradually as more and more of the helix becomes disordered in smaller steps. Since the calculation includes an estimate of starting velocity inferred from simple non-denaturing gels, we may expect that the time scale suggested by the abscissa may not be too far from the real experimental time. In the simple form shown in Figure 9, our experiments provide gel positions for a number of fragments at a single experimental time.

We can ask whether the experimental position and order at that time agrees with the calculated depth and order. There is a difficulty. Although our study has been carried out using fragmented DNA isolated from HeLa cells and the lengths of these fragments correspond to those expected on the basis of the Cambridge sequence[10] derived from the mitochondria of a single human placenta, we have no further assurance that the Cambridge sequence is applicable. The vertical line in Figure 9 represents the time at which the experimental measurements were made; the intersections of each plot with the line represent the expected relative position. These positions are compared with the experimental measurements in Figure 10; the agreement is clear and provides encouragement that the hypotheses relating mobility and melting are on the right track. However, we have compared data from the HeLa DNA derived from an American tumor cell, maintained many years in a laboratory culture, with a recent base sequence from the normal tissue of an English baby. In view of the sensitivity of the experimental system described above, we infer that no substantial variations have accumulated in the HeLa mitochondrial genome.

How can small changes be recognized if they are elsewhere in the fragment than in the lowest melting region? Obviously, for some sequences it will be possible to select different fragment boundaries corresponding to the specificity of a different restriction endonuclease such that a region which might have been second in melting priority according to one cutting scheme becomes first in the other. If the region of interest is further down the priority scale, the chances of finding an appropriate enzyme become vanishingly small. Where the region of interest is low melting but inaccessible because of the absence of appropriate restriction sites, random fragmentation may be useful. With both *lambda* and pBR 322 DNA, it results in well defined, sharply focused zones, and we think that resolution will be adequate for detecting many small alterations. Modification of the denaturing environment of the gel by the incorporation of substances with selected binding to either the double helical or the disordered form of DNA may change the melting map and result in a reordering of melting priority. The sequences that do not determine retardation in the simple system will become visible, augmenting the fraction of the genome recognizable by direct examination. Altogether, direct examination of restriction or random fragments can be expected to permit recognition of small changes within a substantial part of the genome, but perhaps less than half.

There is a more elaborate tactic by which virtually any sequence in the human genome can be transformed into the retardation determinant. Using recombinant techniques, the sequence of interest can be attached at one or both ends to a higher melting sequence of appropriate length. Since the difference in melting temperatures of the test and attachment sequences need not be large, only rare, extraordinarily GC-dense sequences could not be accommodated. The details of the high-melting attachment sequences are not significant so long as they remain stable through cultivation as recombinants, and the same attachments can be used for a great variety of test sequences, since they will not contribute to the earliest melting. We have adopted this approach in collaboration with T. Maniatis and R. Myers to search for mutations in the promotor region of β-globin. Experiments now in progress have been designed according to theoretical calculations on the effect of the attachment sequences on the melting map, the retardation characteristics, and the effect of base substitutions.

In the upper panel of Figure 11, the melting map for a section of chromosome containing the β-globin promotor is shown, and the panels below give melting maps of the same sequence with a high melting tail attached to the right or left end. The tail sequence was derived by Maniatis and Myers from the human α-globin promotor. Although it is extremely dense in guanine and cytosine, it contains enough sequence variety to be propagated through many generations in a recombinant plasmid. With an appropriate choice of linkers and restriction sites, the promotor sequence and the attached tail can be isolated as a single fragment from the plasmid DNA within which they are grown. It can be seen from the top panel of Figure 11 that the promotor sequence is not recognizable as a low melting domain in its original context. The uniformity of the melting temperature for the entire fragment indicates that its strands would be expected to dissociate abruptly and completely in the gel without substantial retardation. The second panel shows the melting map of the short promotor fragment alone. The prospects are radically changed by the attachment of a high-melting tail. With the tail on the left, the promotor becomes effectively a single domain with a very small, slightly lower melting shoulder, shown in panel c. It can be seen in panel d that attachment of the same tail to the other end of the promotor sequence is better; the promotor becomes one sharp domain.

Since the recombinant molecule with a high melting tail at the left end seems satisfactory and is convenient to prepare, we can explore the expected effects of single base changes. Perturbations of the melting map can be presented more clearly as a difference between the melting temperatures of the original and the altered sequence as in Figure 5, rather than in terms of a new map over the full temperature scale. Figure 12 shows the difference between the original melting map for the recombinant with a left end tail and the maps calculated for a base substitution between the same pair of neighboring bases at positions 317, 358, 398 and 432. In each of these four, a triplet TAC has been replaced by the triplet TGC. Differing from the effects in the *lambda cy* fragment, these substitutions effect a somewhat more local change in the melting pattern, rather than an elevation or depression of the entire domain. The effect is distributed by the cooperativity over 20 to 50 bases, not the entire 200. Calculation of the travel schedule in the same

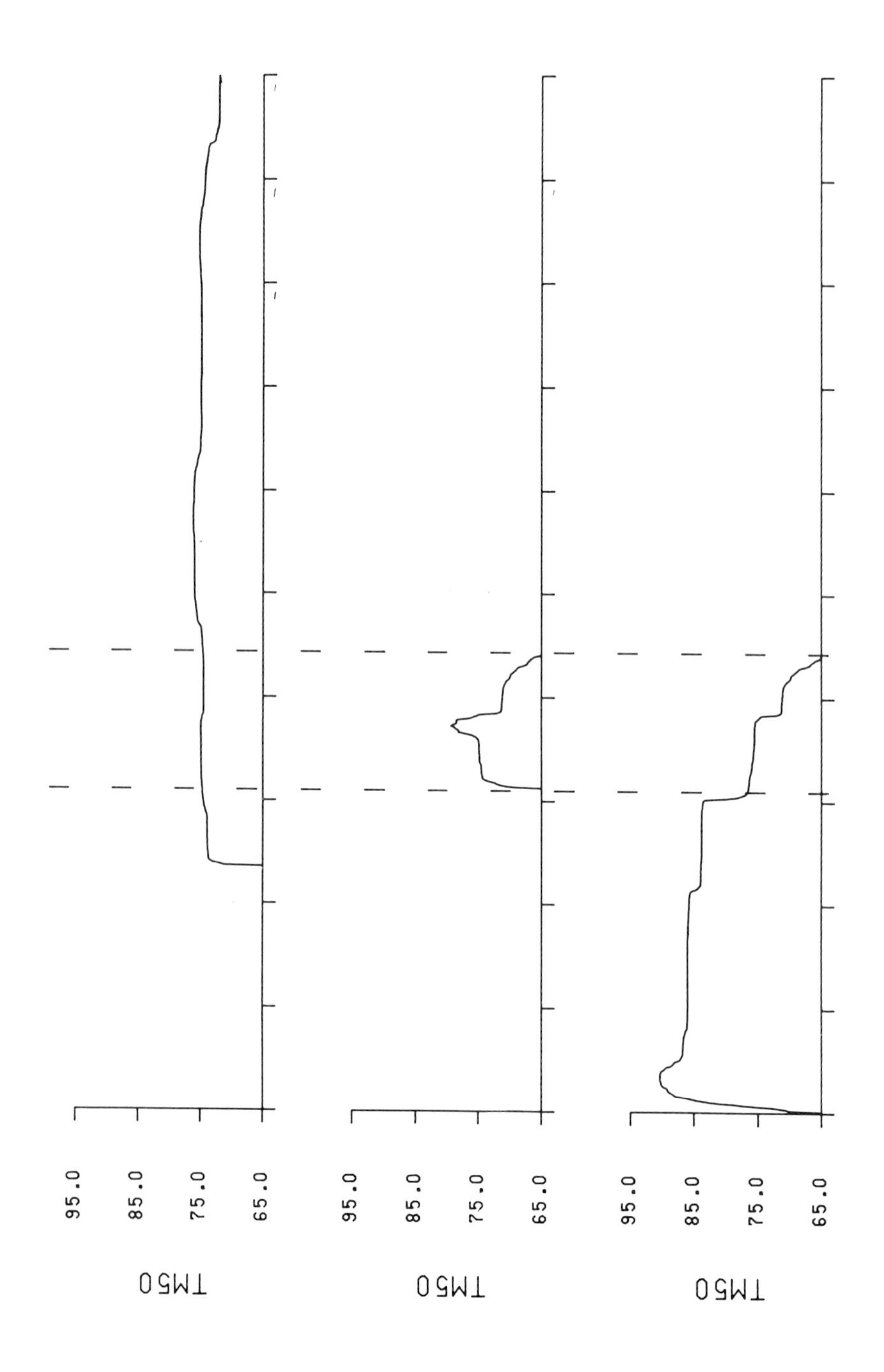
95.0
85.0
75.0
65.0
TM50
95.0
85.0
75.0
65.0
TM50
95.0
85.0
75.0
65.0
TM50

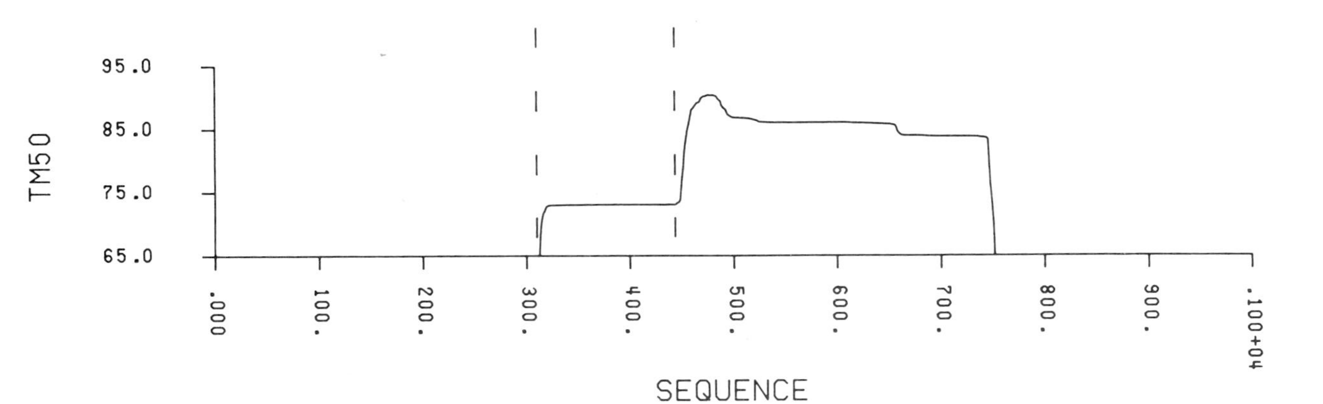

Fig. 11. Melting maps for a short segment of the mouse major β-globin promotor and hypothetical recombinants.
A (top): Melting map of a longer sequence containing the full promotor.
B: Melting map of the short sequence of interest.
C: Melting map with the same short mouse sequence linked at its left end to a longer GC-dense segment derived from the human α-globin promoter.
D: (bottom): The same short mouse sequence linked to the same GC-dense segment at its right end.

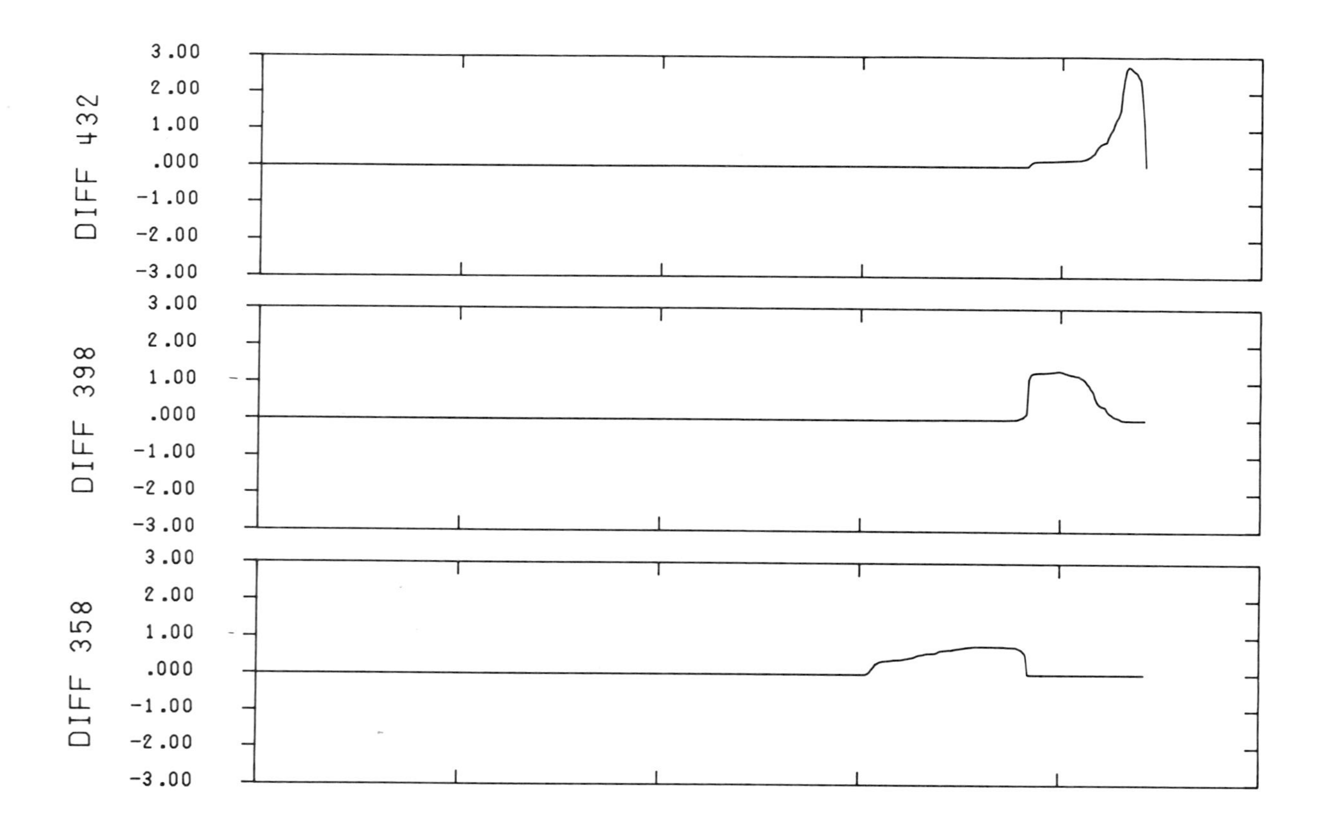

DIFF 432
3.00
2.00
1.00
.000
-1.00
-2.00
-3.00
DIFF 398
3.00
2.00
1.00
.000
-1.00
-2.00
-3.00
DIFF 358
3.00
2.00
1.00
.000
-1.00
-2.00
-3.00

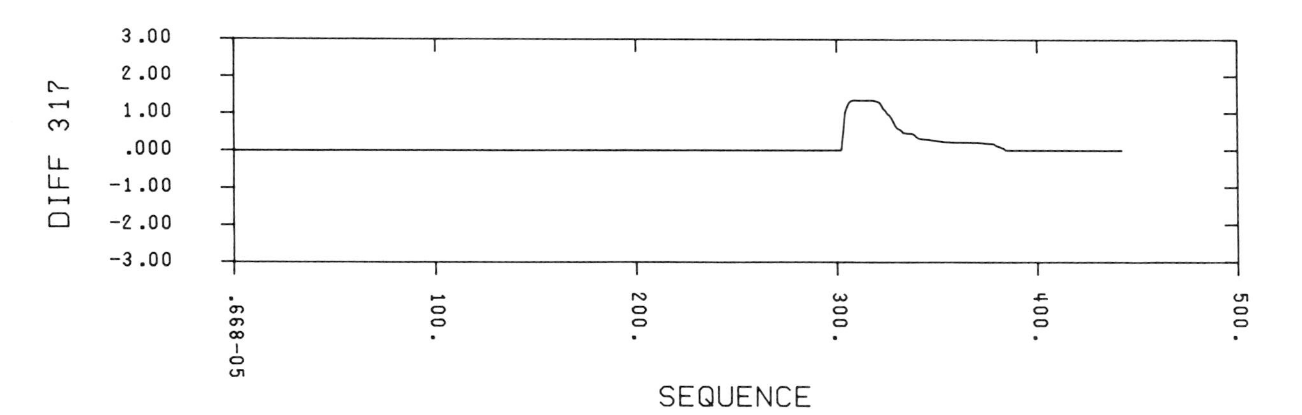

Fig. 12. Difference melting maps for the replacement of an AT pair by GC at four different sites in the mouse major β-globin promotor recombinant. In each of the sites, the original sequence is TAG.

way as that shown for the mitochondrial fragments in Figure 9 leads to the expectation that these substitutions will result in gel displacements of 2 to 5 mm.

If the fragments are isolated from cells that have been subjected to rigorous mutagenesis, the occasional sequence alterations will result in a broader distribution of DNA near the normal retardation position. While the great majority of fragment molecules will still concentrate in the normal band, those that have undergone substitution or other alterations should be found either above or below the band. If DNA can be collected from the gel away from the main band and cloned again for amplification, it may be possible to achieve a very large, non-selective enrichment for a variety of sequence alterations. In the same sense, naturally occurring mutations in a homogeneous molecular population should be easy to perceive.

Insertion or deletion of one or two bases in a sequence also leads to significant changes in the expected melting pattern, and we expect that they will be detectable in the gel. Figure 13 shows the travel schedule calculated for the tailed mouse β-globin promotor and the effects of a deletion of one of three adjacent G's at 376–378 and the effect of an adjacent duplication of the AT doublet at 141–143. In the denaturing gradient gel of Figure 3, the displacement of the mutant band from the wild type would be expected to be about 2 cm.

Much of our discussion up to now has centered on detecting relatively small variations in genotype, in part because these are the changes we associate with most of the well-understood metabolic errors and, in part because they represent the most difficult challenge. Aside from the presence of adventitious sequence elements, such as viruses and plasmids, major sequence aberrations occur at the site of transpositions and transposms. We have shown that restriction endonuclease fragments can be closely characterized in a two-dimensional gel system where the DNA sample is first segregated according to length by conventional gel electrophoresis, and then sorted in the perpendicular direction in a denaturing gradient.[9] Upwards of a thousand fragments can be displayed in a two-dimensional coordinate system. A sequence aberration would be demonstrable either as a perturbation in depth of a fragment of unchanged length, or by more complex patterns of changes. Even the simplest disappearance of a restriction site can be detectable in this system, and not otherwise in a genome the size of a bacterium where the large multiplicity of endonuclease fragments renders an ordinary one-dimensional length separation an unintelligible blur in which a single change is imperceptible. Figure 14 shows a section of a two-dimensional pattern prepared in this way from the DNA of *E. coli* carrying an integrated *lambda* prophage. Several spots that are indicated in this pattern are not found in the pattern below given by DNA of the corresponding *coli* strain lacking the prophage. The pattern provides a substantial characterization of a new element inserted into the *coli* genome, an increment of only 1% of the original. Although the appeal of applying a procedure like this to identify fragments of the human genome is obvious, it withers in the arithmetic of the system; fragmentation of the human genome with a highly-specific restriction

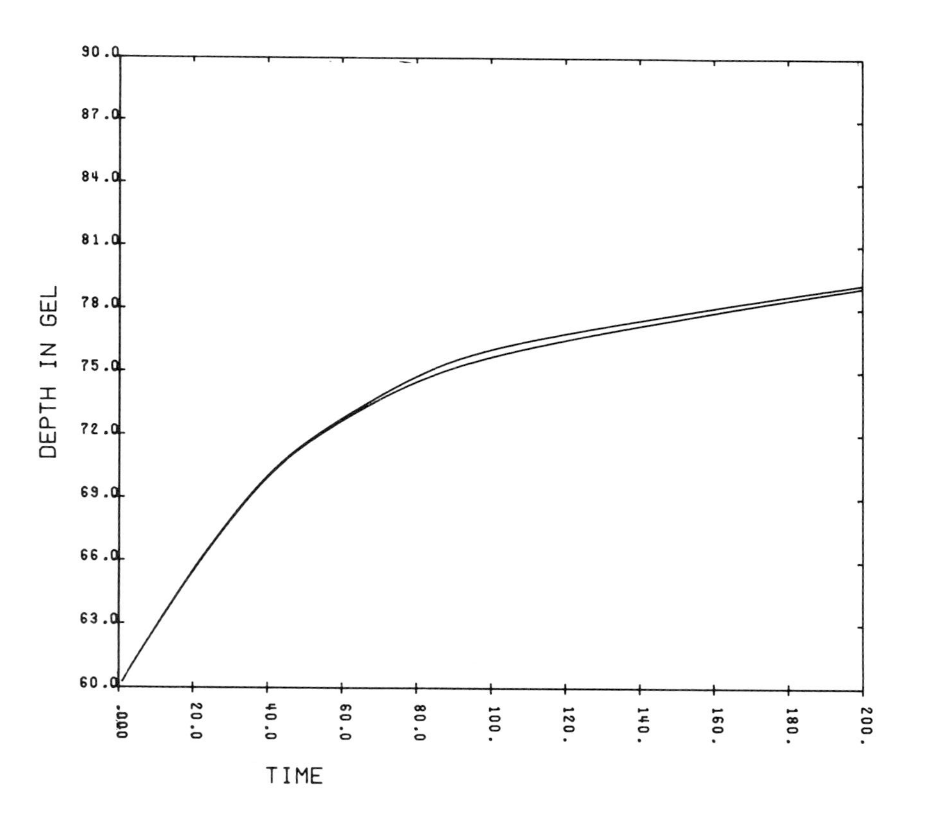

Fig. 13. The change in the calculated movement of a recombinant promotor fragment in a denaturing gradient gel as the result of the deletion of three adjacent GC pairs. The curve showing deeper penetration into the gradient was calculated for the normal mouse sequence.

Fig. 14. Two-dimensional separation of *Eco*RI restriction fragments of the DNA of *E. coli* and an isogenic lysogenic strain carrying an integrated λ prophage. Length separation is right to left, and gradient separation is top to bottom, as in Figure 7. Fragments from both sources were present in the same gel with a very much smaller amount of the lysogenic DNA, ^{32}p-labelled, and no label in the DNA of the parental *coli* strain. The ethidium fluorescence (upper panel) has a negligible contribution from the lysogenic strain, which is solely represented in the autoradiogram. Several spots attributable to the integrated genome are visible in the lower pattern, but absent in the upper. Taken from ref. 9, but only part of the gel is shown.

endonuclease such as *Eco*RII can be expected to provide on the order of a million fragments. Even with the impressive resolution of the two-dimensional gels, three million spots spread over a six inch square represent an unresolvable blur. However, if a probe is available for a specific fragment of interest, the same procedures that are applicable to low-resolution, one-dimensional electrophoresis can be applied, retaining the benefits of two-dimensional dispersion. After two-dimensional separation, DNA can be transferred to an appropriate membrane, in which the locus of the fragments of interest can be determined by hybridization with a radioactive or fluorescent probe.

This work offers new techniques for genomic analysis that may have diverse applications. It also provides a close scrutiny of the relation between sequence and physical properties of the double helix. It represents a novel challenge to theoretical understanding of the DNA molecule.

SUMMARY

We may consider the following points:

1. By application of newly-recognized physical properties of DNA, we have been able to separate DNA molecules according to the characteristics of base sequence.
2. The separation is sufficiently sensitive to recognize alterations that are, in principle, the most minute genetic changes.
3. The laboratory operations required are rapid and simple, suitable for routine screening.
4. The physical separation of differing molecules permits small-scale preparative isolation or enrichment and spatial discrimination by less specific, hybridizing probes.
5. The theoretical foundation for the separations provides major guidance in the design of the experiments and permits well-defined inferences on the nature of the sequences.
6. The detailed characterization of DNA molecules provided by a combination of the sequence-determined separation with length-determined separation in a two-dimensional system permits recognition of otherwise obscure changes; and comprehensive, coarse scale mapping of very long sequences.

ACKNOWLEDGMENTS

This work was supported by grant GM-24030-06 from the National Institutes of Health. We are grateful to the New York State Department of Health Research Council for their grant C175293, to Karen Silverstein and Nelly Brown for their assistance, and to Ryland Loos for executing the drawing used for Figure 1.

Figures 3, 4, 5, and 6, have been reproduced by courtesy of the National Academy of Sciences; Figure 7 by courtesy of Adenine Press.

REFERENCES

1. Schroedinger E. What Is Life? Cambridge, England:Cambridge University Press, 1945.
2. Poland D. Recursion relation generation of probability profiles for specific sequence macromolecules with long-range correlations. *Biopolymers* 1974; 13: 1859-71.
3. Fixman M, Friere JJ. Theory of DNA melting curves. *Biopolymers* 1977; 16:2693-2704.
4. Gotoh O, Tagashira Y. Stabilities of nearest-neighbor doublets in double helical DNA determined by fitting calculated melting profiles to observed profiles. *Biopolymers* 1981; 20: 1033-43.
5. Fischer SG, Lerman LS. Separation of random fragments of DNA according to properties of their sequences. *Proc Nat Acad Sci USA* 1980; 77: 4420-24.
6. Fischer SG, Lerman LS. DNA fragments differing by single base substitutions are separated in denaturing gradient gels in correspondence with melting theory. *Proc Nat Acad Sci USA* 1983; in press.
7. Wulff DL, *et al.* Structure and function of the *cy* control region of bacteriophage *lambda. J Molec Biol* 1980; 138:209-30.
8. Lerman LS, Fischer SG, Bregman DB, Silverstein KJ. Base sequence and melting thermodynamics determine the position of pBR322 fragments in two-dimensional gel electrophoresis. *In:* Sarma RH, ed. Biomolecular Stereodynamics. New York:Adenine Press, 1981: 459-70.
9. Fischer SG, Lerman LS. Length-independent separation of DNA restriction fragments in two-dimensional gel electrophoresis. *Cell* 1979; 16:191-200.
10. Anderson S, *et al.* Sequence and organization of the human mitochondrial genome. *Nature* 1981; 290:457-65.
11. Lerman LS, Frisch HL. Why does the electrophoretic mobility of DNA in gels vary with the length of the molecule? *Biopolymers* 1982; 21:995-97.

ABSTRACTS

DISCRIMINATION OF SINGLE BASE PAIR SUBSTITUTIONS IN DENATURING GRADIENT GELS AND IN THEORETICAL MELTING MAPS

Stuart G. Fischer
Leonard S. Lerman

Nearly identical DNA fragments of 536 base pairs in length differing by a single AT to GC, GC to AT, GC to CG or AT to TA substitution were resolved by denaturing gradient electrophoresis. Separation occurs when the substitution alters the denaturant concentration at which the mobility of the DNA molecule is abruptly retarded, determining the final depth into the gel. Theoretical, sequence-specific melting analysis shows that mutations affecting final gel penetration all lie within a 120 base pair region, the first long sequence in the fragment to melt cooperatively. Substitutions in an adjacent, higher melting domain establish final gel positions indistinguishable from wild type.

REFERENCES

1. Fischer SG, Lerman LS. Separation of random fragments of DNA according to properties of their sequences. *Proc Natl Acad Sci USA* 1980; 77:4420-24.
2. Lerman LS, Fischer SG, Bregman DB, Silverstein KJ. Base sequence and melting thermodynamics determine the position of pBR322 fragments in two-dimensional gel electrophoresis. *In:* Sarma RH, ed. Biomolecular Stereodynamics. New York:Adenine Press, 1981:459-70.

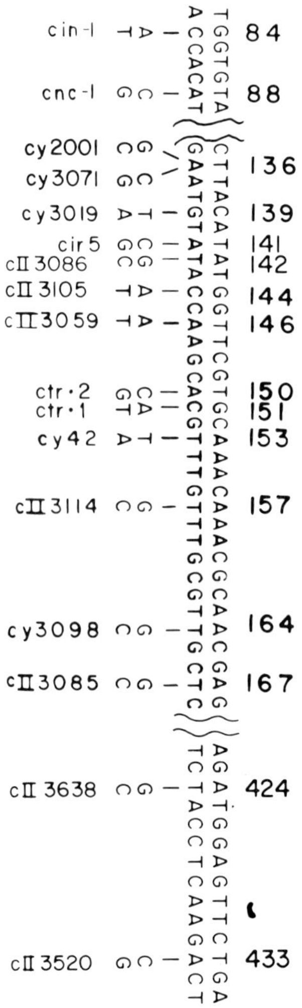

cin-1 84
cnc-1 88
cy2001 136
cy3071
cy3019 139
cir5 141
cII 3086 142
cII 3105 144
cII 3059 146
ctr·2 150
ctr·1 151
cy42 153
cII 3114 157
cy3098 164
cII 3085 167
cII 3638 424
cII 3520 433

SITES OF SUBSTITUTION

DIRECT GENE DOSAGE DETERMINATION IN PATIENTS WITH UNBALANCED CHROMOSOMAL ABERRATIONS USING CLONED DNA SEQUENCES

Claudine Junien
Cecile Huerre
Marie-Odile Rethore

Gene dosage studies in patients with chromosomal aberrations have provided the most precise regional chromosomal assignments for genes expressed in red and white blood cells and in fibroblasts. We now report the feasibility of direct gene dosage using a cloned DNA segment as a molecular hybridization probe. Two different approaches are described. (1) The intensity of the signal obtained by hybridization of the radioactive probe to the corresponding DNA fragment can be compared with the intensity of the DNA fragments which hybridize with a non-syntenic probe used as an internal control. (2) The probe can detect a RFLP: if both parents have different alleles it can be established whether the child is monosomic, disomic or trisomic for this locus.

This has been demonstrated by densitometer tracing of the autoradiogram using an X-specific DNA sequence (see figure), β-globin and $\alpha 2$(I) collagen, in normal men and women, in one patient trisomic for 11p, and in one patient trisomic for segment 7q21→7qter. The man/woman ratio for the X-specific sequence was close to the expected value 0.5 (see figure) while the trisomic patient/normal control ratio was close to 1.5 for β-globin and $\alpha 2$(I) collagen (data not shown). The gene coding for $\alpha 2$(I) collagen can therefore be assigned to 7q21→7qter. This method should also apply to non coding sequences. The increasing number of cloned DNA segments that have already been assigned to a specific chromosome represent a new tool for prenatal and premorbid diagnosis of unbalanced chromosomal aberrations. Complete data are in press.[1]

REFERENCE

1. Junien C, Huerre C, Rethore MO. Direct gene dosage determination in patients with unbalanced chromosomal aberrations using cloned DNA sequences. Application to the regional assignment of the gene for $\alpha 2$(I) procollagen (COLIA2). *Am J Hum Genet* 1983 in press.

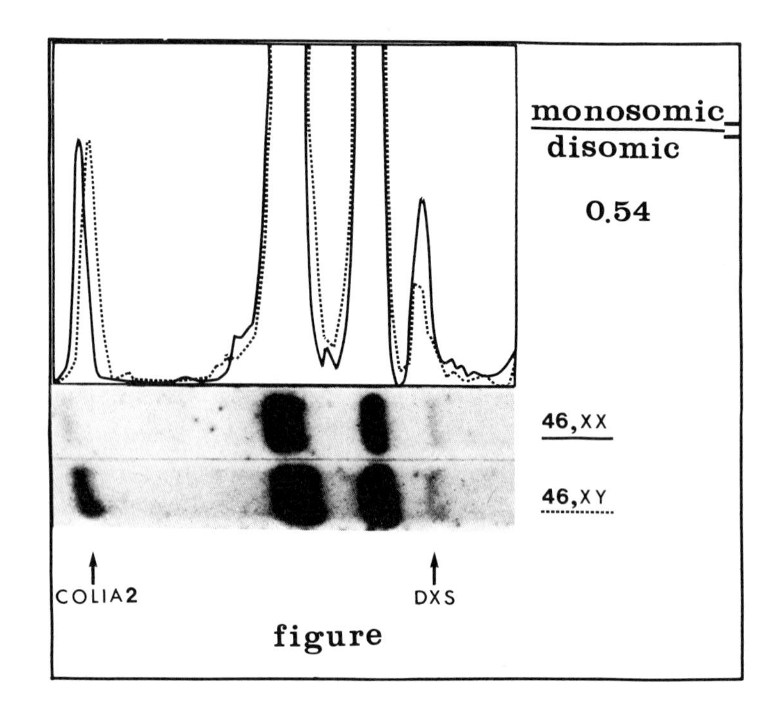

ANALYSIS OF DENSITOMETER SCANNING PROFILES

The intensity of the signals is evaluated by densitometer scanning of the autoradiograms (Gelman apparatus). The area of the peaks corresponding to each fragment is computed. The average peak area of the fragments is calculated by analysis of several traces in several lanes corresponding to the same individual. Calculation of the mean peak area ratio of the probe to the internal control: the height of the peak used as an internal standard (*i.e.* COLIA2) was adjusted to the same level in the different lanes to be compared by expanding the scale sensitivity of the Gelman apparatus. The areas under the peaks were calculated in several traces and in several lanes for the same individual.

MAP OF PORTIONS OF THE HUMAN GENOME PRESENT IN SINGLE OR TRIPLE DOSE IN CELL CULTURES AVAILABLE FROM THE NIGMS HUMAN GENETIC MUTANT CELL REPOSITORY

Margaret M. Aronson
Warren W. Nichols
Richard A. Mulivor
Arthur E. Greene
Lewis L. Coriell

Stored human cell cultures derived from individuals with unbalanced chromosomal aberrations and their carrier parents have played a major role in gene mapping by the methods of somatic cell hybridization. Most of the structural genes mapped have been localized in this way, which requires that a detectable human gene product be expressed in the hybrid cells. Direct mapping of a specific gene sequence is now possible using recombinant DNA technology to locate a cloned specific DNA probe by Southern blot analysis of human-rodent hybrid cell DNA or of flow sorted chromosomal DNA. Cell cultures are being used in exploiting the methods of molecular mapping (*e.g.* in the cloning of a chromosome-specific library from a cell line with multiple copies of a single chromosome[1] or chromosomal region.) In regional localization of a specific DNA probe to a chromosomal segment by *in situ* hybridization, the human hybrid parent cells came from balanced translocation cultures with different breakpoints in that region.[2]

To aid in the identification of cell cultures which have portions of the genome present in single or triple dose, we have prepared a map of available cultures by chromosomal location. The accompanying figure shows cultures by Repository number which are monosomic or trisomic for portions of chromosome 16. A complete set of chromosomal maps has been submitted for publication in *Cytogenetics and Cell Genetics* (1983).

REFERENCES

1. Davies KE, Young BD, Elles RG, Hill ME, Williamson R. Cloning of a representative genomic library of the human X chromosome after sorting by flow cytometry. *Nature* 1981; 293:374-76.
2. Barg R, *et al.* Regional localization of the human globin gene to the short arm of chromosome 16 (16p-pter) using both somatic cell hybrids and *in situ* hybridization. *Gene Mapping* 1982; 6:252.

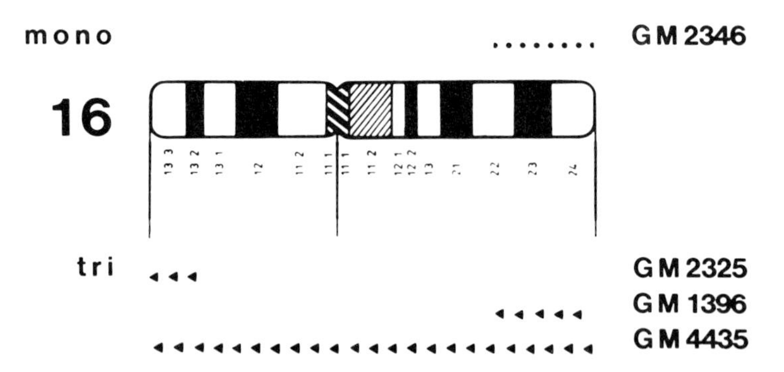

A MAP OF PORTIONS OF CHROMOSOME 16 PRESENT IN SINGLE OR TRIPLE DOSE IN SPECIFIC MUTANT CELL CULTURES

The extent of each monosomic (above the diagram) or trisomic (below) region is indicated relative to the 14 G bands on a standard ideogram of chromosome 16 (ISCN 1981, 550 bands). For example, the fibroblast culture, GM 2346 is monosomic for the region 16q22 to 16qter. GM 4435, a culture established from a spontaneous abortion is trisomic for the entire chromosome.

CLINICAL USE OF REPETITIVE DNA

Robert D. Burk
Judy Stamberg
Keith E. Young
Kirby D. Smith

Recombinant DNA technology has made available a variety of specific gene probes for molecular diagnoses. However, a large fraction of the genome is devoid either of known genes or of genes for which specific probes exist. Some such genomic regions may be identified by specific sets of repeated DNA elements. The use of repeated DNA offers the additional advantage that they may be used for cytologic as well as molecular analyses. We have used two repeated DNA fragments (3.4 and 2.1 kb) released from Y chromosome DNA by digestion with the restriction endonuclease *Hae* III to analyze potential Y chromosome/autosome translocations. In one case, a phenotypically normal female was referred to us for a chromosome study. Her daughter was diagnosed at another institution to have a balanced 9/22 chromosomal translocation. She had a therapeutic abortion when chromosome analysis following amniocentesis indicated her fetus had an unbalanced 9/22 translocation. Detailed chromosome analyses of the proposita performed here by G, Q, and C banding and NOR silver staining revealed the extra material on chromosome 22 to be heterochromatic and thus likely to be either an enlarged satellite or a translocation of the distal Yq chromosome arm. Analysis of the proposita DNA with Y specific repeated DNA probes revealed the presence of both the 3.4 and 2.1 kb male (Y) specific fragments. Thus, the additional fluorescent chromosomal segment carried by chromosome 22 in this individual is most likely from the Y. In the second case, also a phenotypically normal female, analysis by G, C, and Q banding and NOR silver staining revealed extra heterochromatic material on chromosome 22. Analysis of her DNA for Y specific repeated DNAs was negative indicating that the extra chromosomal segment was not from the Y chromosome. These two cases demonstrate that repeated DNA can distinguish some chromosomal translocations and may be useful in karyotypic and prenatal diagnoses.

MAPPING DNA RESTRICTION FRAGMENT POLYMORPHISMS IN THE MOUSE

Rosemary W. Elliott
Elizabeth Mann
Beverly Richards-Smith

In a search for DNA-level polymorphisms, we have screened endonuclease-restricted DNA from eight inbred strains of Mus musculus and from individual animals of the wild species Mus spretus, using 30 cloned cDNA probes. We found two to be mitochondrial and the rest nuclear.

We have defined twenty-four genetic loci in M. musculus, of which thirteen have been mapped, with one site on each of chromosomes 1, 3, 4, 12, 14 and 17, two on chromosome 6 and five on chromosome 7. For every clone studied there is a difference in at least one hybridizing fragment between M. musculus and M. spretus.

We present here data for p1109 which defines the polymorphism on chromosome 17 and for p1443 which defines one of the polymorphisms on chromosome 7. Both hybridize with single DNA fragments produced by restriction enzymes BamHI, EcoRI and HindIII.

Three phenotypes were found among the eight strains when p1109 was hybridized to Southern blots of mouse DNA digested with BamHI and separated by electrophoresis. Using the recombinant inbred AKXL and SWXL strains we have shown that the fragments are probably allelic and map on chromosome 17, distal to the H-2 locus.

Only one variant strain was found when p1443 was hybridized to Southern blots of *BamHI*-treated and electrophoretically separated mouse DNA. Using the recombinant inbred BXD strains we have shown that the fragments map on chromosome 7 between *Gpi* and *Tam*-1.

REFERENCES

1. Elliott RW. Use of two-dimensional electrophoresis to identify and map new mouse genes. *Genetics* 1979; 91:295-308.
2. Elliott RW, Berger FG. DNA sequence polymorphism in an androgen-regulated gene is associated with alteration in the encoded RNAs. *Proc Natl Acad Sci USA* 1983; 80:501-4.

METHYLATION PATTERNS OF THE MAJOR AND MINOR SATELLITE SEQUENCES IN THE GERM CELLS OF *MUS MUSCULUS*

Janet P. Sanford
Lesley M. Forrester
Verne M. Chapman
Nicholas D. Hastie

Recent studies suggest a correlation between hypomethylated DNA and active genes. Using *HpaII/MspI* restriction endonucleases and Southern blot hybridization, we have examined methylation patterns of the major and minor satellite sequences in germ cells and germ cell precursors of *Mus musculus*.

Our laboratory constructed a repetitive DNA library from *Mus musculus* and isolated two families of satellite DNA sequences called major and minor. Both map to centromeric regions based on *in situ* hybridization studies. The major and minor sequences are present in $>10^6$ and 75,000 copies per haploid genome, respectively.

Our studies show that satellite DNA in somatic tissues such as liver are highly methylated. In contrast, sperm and oocyte satellite sequences are hypomethylated relative to the somatic tissues. This is evident for both the major and the minor satellites.

We have examined the methylation patterns of the minor satellite during germ cell differentiation. Cell populations were isolated from testes tubules of neonatal mice at various stages, starting at one day after birth through the eighth day. The germ cell population in one-day old testes was predominantly gonocytes. The transition to spermatogonia was observed by day three. The *HpaII/MspI* digestion patterns of the minor satellite in gonocytes were similar to that seen in somatic tissues. *HpaII* digestion of minor satellite was first observed in day three germ cells, and by day five, the pattern was similar to that observed in mature sperm. The results indicate that the minor satellite sequences become demethylated during germ cell differentiation, coincident with the transition from gonocyte to spermatogonia.

p150
Sperm Oocyte Liver
H M H M H M H
Methylation
of Minor Satellite DNA
in Various Tissues

HYPOMETHYLATION OF REPETITIVE ELEMENTS IN EXTRAEMBRYONIC TISSUES OF MICE

Lesley M. Forrester
Verne M. Chapman
Nicholas D. Hastie
Janet Rossant

Several families of highly- and middle-repetitive sequences have been identified in the mouse genome from a clone library of mouse DNA. These sequences include centromerically associated satellite DNA as well as DNA which is dispersed throughout the mouse genome. Preliminary evidence indicates that these sequences are highly methylated in somatic tissues of the mouse but they undergo at least partial demethylation in germ cells during the process of gametogenesis and show some degree of hypomethylation in mature sperm and oocytes (Sanford *et al*, these proceedings).

Studies of DNA methylation in rabbit trophoblast suggest that these extraembryonic derivatives are hypomethylated compared to embryoblast cells, but no specific sequences were identified (Manes and Menzel, 1981). We have examined the patterns of methylation in different extraembryonic derivatives and the embryo proper to determine whether specific mouse repetitive sequences, which are hypomethylated in germ cells, are also undermethylated in extraembryonic cells. We used Southern blot analysis of DNA digested with *MspI* and *HpaII* to identify changes in DNA methylation at CpG sites. We specifically examined changes in methylation of a minor satellite sequence identified by probe pMR150 and a dispersed repetitive element *MIF* identified by the probe pMR134. *MspI* digestion of both of these sequences produces DNA fragments which are readily identified in Southern blot analysis. We observed that the pattern of *HpaII* digestion fragments for the minor satellite in two extraembryonic derivative tissues, ectoplacental cone from the trophoblast and yolk sac endoderm from primitive endoderm, was similar to that observed in mature germ cells. By contrast, no *HpaII* digestion fragments could be detected in the egg cylinder (embryonic) DNA at the same stage. These results indicate that these repetitive undermethylated elements become remethylated in the embryonic lineages but that these sequences are not completely methylated in the extraembryonic lineages of trophectoderm and yolk sac endoderm.

GENETIC AND MOLECULAR ANALYSIS OF MOUSE SUBMAXILLARY GLAND RENIN

Nina Piccini
Douglas Dickinson
John L. Knopf
Kenneth Abel
Kenneth W. Gross

The renin regulatory locus (Rnr) is a genetic element governing mouse submaxillary gland (SMG) renin levels.[1] A 45,000 dalton polypeptide detectable after *in vitro* translation of mouse SMG mRNA has been identified by genetic and physical criteria as SMG renin. A cDNA recombinant clone specific for SMG renin has been isolated and used to demonstrate that the previously described genetic regulation of SMG renin levels is manifested at the level of renin mRNA concentration. The renin cDNA clone has also been used in Southern blot analyses to study the organization of homologous DNA sequences in strains carrying different alleles at the Rnr locus. Restriction digest patterns of high renin strains (Rnr^s) are characteristically distinct from patterns observed for low renin strains (Rnr^b) and suggestive of a structural gene duplication at the chromosome 1 locus in high renin strains.[2] However, gene dosage cannot account for the increased renin levels in high renin strains since Rnr^s and Rnr^b strains may differ up to 100 fold in SMG renin levels. Finally, variants for renin structural gene organization and expression have been identified within wild mouse populations.

REFERENCES

1. Wilson CM, Erdos EG, Wilson JD, Taylor BA. Location on chromosome 1 of Rnr, a gene that regulates renin in the submaxillary gland of the mouse. *Proc Natl Acad Sci. USA* 1978; 75:5623-26.
2. Piccini N, Knopf JL, Gross KW. A DNA polymorphism, consistent with gene duplication, correlates with high renin levels in the mouse submaxillary gland. *Cell* 1982; 30:205-13.

ISOLATION AND CHARACTERIZATION OF A CLONED cDNA COMPLEMENTARY TO MURINE GLUCURONIDASE mRNA

Robbin Palmer
Patricia M. Gallagher
William L. Boyko
Dian E. Grogan
Roger E. Ganschow

Studies from our laboratory and by others have described and analyzed two genetic regulatory elements, designated *Gus-r* and *Gus-t,* which control the rate of murine glucuronidase (GUS) synthesis. Both elements are tightly linked to the GUS structural gene (*Gus-s)* on chromosome 5. *Gus-r* is a cis-active regulator of the androgen-induced rates of GUS synthesis in kidney tubule cells while *Gus-t* is a *trans*-active regulator of the basal rates of GUS synthesis in all tissues, with significant tissue-specific effects during post-natal development. In order to analyze these regulatory elements at the nucleic acid level, we have developed a cloned cDNA complementary to murine GUS mRNA.

Double-stranded cDNA was synthesized using androgen-stimulated mouse kidney poly(A) + RNA which had been enriched for GUS mRNA activity by sucrose gradient sedimentation. When this cDNA preparation was inserted into the *PstI* site of pBR322 the resulting recombinant DNA generated 3000 transformants upon transfer into *E. coli.* This cDNA library was screened for GUS cDNA by binding DNA from pools of 12 clones to nitrocellulose filters and testing the ability of each filter to hybridize GUS mRNA. Filter-selected mRNA's were assayed for their ability to program the incorporation of radioactive amino acids into immunoprecipitable GUS when added to a modified reticulocyte lysate. Two positive pools were found among 960 tested clones and the individual clones of one of these pools were tested using the same protocol. A positive clone containing an insert of approximately 1.8 kilobases was identified and tested for its ability to hybridization-select GUS mRNA from total mouse kidney RNA preparations which varied in GUS mRNA activity as a function of both hormonal and genetic differences. Relative amounts of GUS mRNA activity selected from each preparation by this DNA were similar to those previously determined by both *in vivo* and *in vitro* assays, providing strong evidence that sequences complementary to murine GUS mRNA are present in this recombinant clone. (Supported by USPHS grant AM-14770.)

REFERENCES

1. Swank RT, Paigen K, Ganschow RE. Genetic control of glucuronidase induction in mice. *J Mol Biol* 1973; 81:225-43.
2. Ganschow RE. Simultaneous genetic control of the structure and rate of synthesis of murine glucuronidase. *IN:* Markert CL, ed. Isonzymes: Genetics and Evolution. Vol. 4. New York: Academic Press, 1975:633-47.

VARIATION IN EXPRESSION AND RANDOM CHROMOSOMAL INTEGRATION OF CLONED HLA GENOMIC SEQUENCES IN MOUSE LTK$^-$ CELLS FOLLOWING DNA–MEDIATED GENE TRANSFER

James A. Barbosa
Michael E. Kamark
A. Ferguson-Smith
Frank H. Ruddle

We have identified human genomic clones capable of directing the synthesis of the major histocompatibility antigens HLA-A2 and HLA-B7 following DNA-mediated gene transfer (DMGT). The surface expression of these human antigens, monitored by indirect immunofluorescence (IIF) and the fluorescence-activated cell sorter (FACS) was examined at 60 hours after transfection and on HAT resistant (HAT^R) populations derived from cotransfer with the herpes virus thymidine kinase (*HSV-TK*) gene. The frequency of cotransfer of HLA and the level of expression in recipient cells was highly dependent on the *HSV-TK* plasmid used in cotransfections. Although transfection with a plasmid containing only the HSV-TK gene gave low TK transfer frequencies, 85-95% of HAT^R cells expressed high levels of HLA. Cotransfections with plasmids containing the polyoma early region in addition to *HSV-TK* genes showed a 10–50X higher frequency of HAT^R colony formation, but resulted in only 25–40% cotransfer frequences and lower HLA expression levels. Plasmids containing both the HLA genomic sequences and the polyoma enhancing sequences have been constructed and are currently being analyzed.

HAT^R clonal populations isolated from cotransfections with HLA genomic clones displayed varying levels of surface expression that roughly correlated with the number of intact donor HLA sequences present in the transformants, although transformants containing 50–100 intact copies had only 6X higher HLA expression as transformants containing 1–5 intact copies. *In situ* hybridization analysis of ten transformants carrying greater than 20 copies of intact HLA genes reveal only one integration site in each HAT^R clone and integration to be randomly distributed among mouse chromosomes.

REFERENCE

1. Barbosa JA, Kamarck ME, Biro PA, Weissman SM, Ruddle FH. Identification of human genomic clones coding the major histocompatibility antigens HLA-A2 and HLA-B7 by DNA-mediated gene transfer. *Proc Natl Acad Sci USA* 1982; 79:6327-31.

ABERRANT mRNA SYNTHESIZED BY THE INTERNAL DELETION MUTANT OF VESICULAR STOMATITIS VIRUS

Ronald C. Herman

The defective interfering particle DI-LT was generated by an internal deletion of the genome of the heat-resistant strain of vesicular stomatitis virus (VSV-HR).[1] This DI particle synthesizes an aberrant glycoprotein messenger RNA *in vivo.*[2] The abnormal mRNA contains a transcript of the partially deleted polymerase gene co-valently linked to the 3′ end of the glycoprotein message. The polyadenylate is located at the 3′ end of the molecule and is most probably encoded by the rem-nant polymerase gene polyadenylation signal. Some stocks of DI-LT were found to synthesize only the aberrant form of the glycoprotein message (G*) while others synthesized both the normal and abnormal forms in the absence of helper virus. This suggests that some stocks of DI-LT contain more than one DI particle posses-sing the ability to synthesize messenger RNA's. The synthesis of the aberrant G* RNA depends on the nature of the helper virus used to propagate the DI particle. Synthesis of G* correlates directly with the presence of the heat-resistant VSV polymerase. Stocks of DI-LT passaged with the heat-resistant helper virus synthe-size G* while stocks passaged with the wild type virus synthesize the normal G message. These results suggest that the regulatory sequences present at the G/L intercistronic boundary are only conditionally functional in the DI particle. The aberrant message may be synthesized because of a failure to terminate transcription at the end of the glycoprotein gene or because of an inability to process an ab-normal polycistronic precursor.

REFERENCES

1. Epstein DA, Herman RC, Chien IM, Lazzarini RA. Defective interfering partical generated by internal relation of the vesicular stomatitis virus. *J Virology* 1980; 33:818-29.
2. Herman RC, Lazzarini RA. Aberrant glycoprotein mRNA synthesized by the internal deletion mutant of vesicular stomatitis virus. *J Virology* 1981; 40:78-86.

AMPLIFICATION OF THE *E. COLI* AND PHAGE T_4 THYMIDYLATE SYNTHETASE GENES AND CHARACTERIZATION OF THEIR PRODUCTS

Marlene Belfort
Alan Moelleken
Gladys Maley
Frank Maley

The thymidylate synthetase (TSase) genes of *E. coli* K12[1] and bacteriophage T_4 [2] were subcloned into plasmid vectors from λ transducing phages generated *in vitro*. After resection of flanking sequences the TSase genes were placed under control of the phage λ p_L promoter contained in expression plasmid pKC30. Amplification of gene expression resulted in 200– to 1,000–fold elevations in the levels of the synthetases. The amplified bands appearing on SDS-polyacrylamide gels were positively identified as the synthetases by forming the appropriate ternary complex of enzyme, [6^3H] FdUMP–5,10–methylenetetrahydrofolate. Consistent with the differential sensitivity of the two enzymes to inhibition by folate analogs, the purified host and viral synthetases reveal striking dissimilarities in their genetic, kinetic and structural properties. Implications of the diverse nature of duplicated viral and host functions are discussed.

REFERENCES

1. Belfort M, Maley GF, Maley F. Characterization of the *E. coli thy*A gene and its amplified thymidylate synthetase product. *Proc Natl Acad Sci USA* 1983; 80:in press.
2. Belfort M, Moelleken A, Maley GF, Maley F. Purification and properties of the phage T_4 thymidylate synthetase produced by the cloned gene in an amplification vector. *J Biol Chem* 1983; 258:in press.

Index